THE SCIENCE OF THE SOUL AND THE STARS

BY

THOMAS H. BURGOYNE

Callender Press 2013

ISBN 9781291516739

CONTENTS

The Science of the Soul and the Stars

Introduction

Chapter I The Zodiac

Chapter II The Constellations

Chapter III The Spiritual Interpretation of the Twelve Houses of the Zodiac

Chapter IV Astro-Theology

Chapter V Astro-Mythology

Chapter VI Symbolism

Chapter VII Alchemy Part I (organic)

Chapter VIII Alchemy Part II (occult)

Chapter IX Talismans

Chapter X Ceremonial Magic

Chapter XI Magic Wands

Chapter XII The Tablets of Aeth in three parts

Part I the Twelve Mansions Part II the Ten Planetary Rulers Part III the Ten Great Powers of the Universe

Chapter XIII Penetralia

PREFACE

No explanation is thought necessary, further than to corroborate the author in all he has said in his somewhat unusual preface.

We have enjoyed, immensely, our work of giving to the world this remarkable series of books on Occultism, and appreciate the large patronage they have received from the reading public, for which we return our sincere thanks. We hope the near future will give us the work referred to by the author in his preface, as doubtless it will be a great revelation of Occult laws that govern our little Earth in its relation to our Sun and solar system, of which it forms a part, and give much light on those subjects that have been shrouded in mystery.

"The Light of Egypt" will be found to be an Occult library in itself, a textbook of esoteric knowledge, setting forth the "wisdom Religion" of life, as taught by the Adepts of Hermetic Philosophy. It will richly repay all who are seeking the higher life to carefully study this book, as it contains in a nutshell the wisdom of the ages regarding man and his destiny, here and hereafter. The London and American first edition, also the French edition, Vol. I, met with lively criticism from Blavatsky Theosophists, because it annihilates that agreeable delusion of "Karma" and "Reincarnation" from the minds of all lovers of truth for truth's sake.

"The Tablets of Aeth" is a great and mighty work, as it contains the very quintessence of Occult and Hermetic philosophy, as revealed by spiritual law. "Penetralia" is a new revelation, and invaluable to Occult students, as it is the personal experience of a developed soul.

To all lovers of Truth we respectfully recommend this Book of Books, as it has justly been called by many who are competent to criticise its teachings. It was the author's wish that his name be withheld from the public, knowing full well that the teachings contained in his works will prove his motto: "Omnia Vincit Veritas."

Now that our author has passed beyond the power of the world to flatter or condemn, and has given his thought for the uplifting of the human family, it is but simple justice that he be made known to the world as its teacher of a higher thought than has preceded him. He shrank from public notoriety, and modestly refused to be publicly known to the world as one of its spiritual leaders for the cycle upon which the Earth and its inhabitants have entered, but the time has come to announce publicly the authorship of the works published anonymously under the symbol of {}, and his writings are to be judged by their merits, and not by prejudice nor personal bias as viewed from the human plane of life.

He moved in the world, comparatively unknown to the world at large, and his greatest friends, though mystified, did not understand his true worth in spiritual greatness. The mask, or person, often hides from view the angel in disguise. Therefore our author must be judged by what he has written, and not by his personators and calumniators. The true student of Occultism always judges the tree by its fruits. If the writings of our author are judged by this standard, they will stand as a beacon light to higher rounds than ours.

These lessons were issued to a few of my pupils as "Private Studies in Occultism," several years ago. The time has now come to give them to the world as a companion to the first volume of "The Light of Egypt."

It is the duty of Occult students to familiarize themselves with the subjects herein discussed. They should know the ideas of our ancestors regarding them and be familiar with their thought, in order to appreciate the sublime wisdom and knowledge of Nature as taught by them, otherwise we are sure to do them, as well as ourselves, great injustice. The history of Occultism bears out the fact that there is very little that is new to the present time.

The arrangement and classification of thought differs during each cycle of time on the different spirals, and, like the fruitage on lower rounds of Nature's progressive wheel of destiny, variety and quality are diverse, so, likewise, do we find the mental manifestations. This age, however, is blessed with a great variety and abundance of thought, in clear-cut language, that should enlighten the races of the Earth with Mother Nature's modus operandi in every department of human thought.

We hope these chapters will aid to this end, and doubtless many students will find in them the key to unlock the mysteries veiled in symbol and hieroglyphic by ancient writers. The author's object has been to make plain and easy of understanding these subjects. Much, however, has been left for private study and research, for many large volumes might be filled if a detailed description of each subject were entered upon, which task is left for those who feel so inclined. A rich reward is in store for those spiritual investigators who will follow out the paths and lines herein mapped out on Spiritual Astrology, Alchemy, and other subjects. Meditation and aspiration will open up hidden treasures that will prove a boon to Occult students, for Astrology and Alchemy are the two grand sciences that explain the why and reasons for what we see and experience on every plane of life. In this age there should be no concealing of these Divine truths. We cannot hide anything in the air, and for this reason the Sun in Aquarius will unearth and reveal to man all that the present cycle has to give during the Sun's passage through this airy sign.

The watery sign, Pisces, through which the Sun manifested during the past 2,160 years, gave up to man their secret powers and hidden attributes in steam as a motive power, which man has completely mastered. He will likewise master the airy forces during the present sub-cycle of the Sun in Aquarius. Already we see him using liquid air and compressed air as a motive power, which will gradually take the place of steam as the Sun gets farther into the sign, or constellation, of Aquarius.

Men will become immensely wiser than they have been, and it is to be hoped they will leave the written record of their achievements in science and art to show to future races their status of mind on every subject for the edification and enlightenment of coming races.

Our ancestors were denied this great privilege. Consequently their wisdom is only symbolized to us in a way that it is difficult to read and interpret correctly, yet we who have the key to their symbols can read accurately the truth they wish to convey, which stands out clearly to all capable of understanding and interpreting symbolism and correspondence correctly. History and Nature repeat themselves in every cycle of time; therefore these forces and potentialities are natural to the sign through which the Sun manifests. We can go backward or forward through the Sun's Zodiac and read correctly the history of the hoary past, as well as the present and future, by bearing in mind the sign and cycle in manifestation at any given period. When the proper time arrives, a work will be given to the world to prove to mankind the law of cycles.

God is present in all ages and races, manifesting His love and wisdom throughout infinite creations, and that He records, in His own way, the most detailed record of any event which takes place, thus giving to man a complete history of His works and will, for man's enlightenment, so that he, too, may cooperate intelligently with his God in every way that intelligence wills to manifest. Prehistoric history is not blotted out from Nature's laboratory. The Astral Book of Karmic evolution will one day reveal its hidden treasures to a waiting world in such a manner as to surprise and enlighten mankind as the recording angels give up those gems of truth they have so jealously guarded for untold cycles of time, simply because the time was not ripe for its divulgence.

There is a time for everything, and when that time arrives all past history of our planet's evolution will be written in an intelligent manner for the illumination and education of man as the masterpiece of the Living God. In this way man will worship Deity and perfect his God-nature, even to Angel-hood.

If this volume of "The Light of Egypt" meets with the same appreciation that was accorded the first volume, which has passed through four editions, and is still growing in favor every day (besides being translated into the French), the author will feel that his efforts have not been wasted, and he trusts the race will have been made better for having read his writings.

As this is his posthumous contribution to the world, the author wishes, in this connection, to pay a debt of gratitude and grateful recognition to his esteemed pupil and friend, Dr. Henry Wagner, who has so generously published nearly all of his writings. Without his aid and assistance we would not have been able, of ourselves, to have given these works to the world. Therefore, honor to whom honor is due.

Mrs. Belle M. Wagner has been chosen by the Masters as my spiritual successor and representative of the

Hermetic Brotherhood of Luxor, and thus perpetuate the chain of outward connection between those in the realm of the higher life with those upon the outward plane.

She is our choice, and a most worthy one to take my place.

I make this statement in this connection for the benefit of my pupils and Hermetic students generally, as I am being personated by frauds and imposters, claiming to be Zanoni. Verbam sap.

It is my request that a fac-simile of my signature and symbol accompany this preface.

Dictated by the author from the subjective plane of life (to which he ascended several years ago) through the law of mental transfer, well known to all Occultists, he is enabled again to speak with those who are still upon the objective plane of life.

The additions found in this volume, not in the original manuscripts, have been supplied in this manner. The two planes of life, the objective and subjective, are scientific facts, no longer disputed by well-informed minds, and the exchange of thought will become almost universal among educated minds during the present cycle. Hence great progress will come to the Earth during the next 2,160 years, while the Sun manifests his glorious influence through the symbol of the Man.

Thanking each and all who have aided in any way to give my writings to the world, I am, in love and fraternal greetings, ever yours. Omnia Vincit Veritas.

INTRODUCTION TO VOL. II

What study is more sublime, inspiring and profitable, in the highest sense, than the "language of the stars"—those silent monitors of the midnight sky, who reveal HIS WILL as secondary causes in the administration of universal law? The science of the stars is the Divine parent of all science.

The more earnest our study, the more recondite our research and thorough our investigation of the "Science of the Stars," the more fully shall we realize the truth of the teacher's words: "Astrology is the key that opens the door to all occult knowledge." It is the key that unlocks the mysteries of man's being; his why, whence, whither. Within the temple of Urania lies concealed the mystery of life. The indices are there, written by the finger of the Infinite in the heavens above.

It is our privilege to make this language our own, and it should be the earnest work of every true student of Nature to acquire a right understanding and correct interpretation of these Divine symbols. And, as thorough students of any language seek out the derivation of words and expressions, search for the root, or stem word, and its origin, so should the student of astrology, by sincere desire and earnest study, seek to know the origin and root of these starry words and complex expressions of the "language of the stars."

The Sun, Moon and five planets[*] of our solar system are to us symbols of the reflected and refracted rays of the triune attributes of the great Central, Spiritual Sun: Life, Light and Love, analogous to the three primary colors in Nature, which become still further refracted into four secondary or complementary colors, rays or attributes, the seven constituting the active principles of Nature, the seven rays of the solar spectrum, the seven notes of a perfect musical scale, there being throughout a perfect correspondence, and all are but different modes of vibration or activities of the Supreme Intelligence.

[*] Uranus and Neptune belonging to a higher octave.

And, as we know, the seven rays of color reflect an almost infinite variety of tints, that, octave upon octave, are built upon the seven natural tones in music, so, also, are these seven active principles divided and subdivided into innumerable forms, qualities and manifestations of the first trinity—Life, Light, Love, life being the manifestation of the second two, love and wisdom, which in turn are the dual expressions of the "One."

Upon the knowledge of these Divine truths Pythagoras built the theory of the "music of the spheres." Let us pause and listen to this celestial music.

Suns and their systems of planets sound forth the deep bass tones and rich tenor, while angelic races take the silvery treble of the Divine melody, octave upon octave, by more and ever more ethereal system upon system, to the very throne of Deity—the Infinite, Eternal source of Light, Life and Love. Let us learn, through the knowledge of the stars, to attune our souls to vibrate to the Divine harmony, so that we may take our places in the celestial choir and blend our voices with those of the celestial singers, chanting the Divine anthem: "We Praise Thee, O God!"

To resume. If we would gain a correct knowledge of astral science we should study astrology in its universal application, side by side with its more intricate phase and the details, as manifested upon the individual man and his material destiny.

Let us digress for a moment. The intellectual minds, the material scientists, who cavil at the "science of the stars," declaring it to be mere fortune-telling, consequently false, do but air their ignorance of this most profound subject, not knowing that it embraces and contains all sciences, all religions, that have ever been or ever will be, comprises all history of every age, of races, empires and nations; that it is the only true chronology, and marks the destiny, not only of personal man on every plane, but of the human family as a whole. All mythologies find their explanation in this starry language, and every religion is founded upon the movements of our solar system. The rise and fall of empires and races of men are written in its pages.

To master as far as we are capable, and our limited space of life here will permit, we must pursue the study in its broad sense, as already stated, in the external application of the starry influx and upon the interior planes of action from God to the mineral, the mineral to man; aye, and man to the angel, finding in every section a complete and perfect correspondence.

To master the alphabet should be the first step, whose vowels, diphthongs and consonants are the planets and shining Zodiac. It is very essential to clearly comprehend the action and reaction of the planets upon the human organism, as an integral part of the universal organism; ever remembering that the starry vowels, in combination with the consonants, or Zodiac, form the infinite expressions comprising the language of the starry heavens in their threefold manifestation upon the external planes of life; while the radiant constellations are the ideas which find expression through this language, which is likewise a science, accurate in its mathematical construction and perfect in geometric proportion.

The student should ever bear in mind that astrology, like every other science, is progressive. The underlying principles are always the same. These are like the "laws of the Medes and Persians," but the plane of action is constantly changing.

It is a well-established and indisputable truth that from the Sun, the solar center of our system, is derived all force, every power and variety of phenomena that manifests itself upon Mother Earth. Therefore, when we remember that the solar parent passes through one sign of his celestial Zodiac in 2,160 years, a twelfth part of his orbit of 25,920 years, we see that from each sign in turn he (the Sun) rays forth an influx peculiar to that special sign; and, as there are no two signs alike in nature or quality, hence the passage of the Sun from one sign into another causes a change of polarity in planetary action, which can be fully demonstrated and conclusively proven. It follows, as a natural sequence, that the rules formulated and taught by astrologers in reference to the plane of planetary influence in one sub-cycle will not hold good in the next. To illustrate: In the year 1881 the Sun passed from the sign Pisces into Aquarius, thus beginning a new cycle of solar force. The human race has entered upon a cycle in every respect differing in nature and action from the past cycle of 2,160 years. The sign Aquarius is masculine, electric, positive. It is intellectual in character, scientific, philosophic, artistic, intuitive and metaphysical. It is the sign of the Man. The truths of the past are becoming etherealized. Our solar parent has scarcely crossed the threshold of the sign Aquarius, and already we observe in many directions the activities of the peculiar influx. True, it is but the first flush of the dawn of a new era, the harbinger of a glorious day to our race. In the light of this truth, ponder well on the nature of the influx radiating from the solar center, each orb of his shining family absorbing a different ray, or attribute, of solar energy, corresponding to its own peculiar nature. The Earth, in her annual passage about her solar parent, receives the harmonious or discordant vibrations of this astral influx according to the many angles she forms to the various planets.

We see, then, that the Earth is enveloped in an atmosphere, or zone, of occult force we recognize as humane, mental, positive, etc., acting and reacting upon the human family through the laws of vibration in strict and exact ratio to its interior capacity to receive and ability to externalize upon the material plane of being. The results, as far as this stage of existence goes, will be manifest as man vibrates harmoniously or otherwise to the stellar cause.

The present sub-cycle producing an entirely different influence to that of the past cycle, whose force was watery, magnetic and feminine, causes a warring of elements, confusion and uncertainty, until the old are displaced by the new conditions. We should learn from these facts that it is folly to brand as false and condemn as worthless the rules and formulas, and even religious thought, of the past when we find upon careful investigation and crucial tests their inadequacy to account for present conditions. They were true in their cycle, and applied to past conditions and states of mental development. But in this new era, upon whose threshold we now stand, the vibrations become more intense. Man's whole nature is being tuned to a higher key. We must not forget that these cycles apply to the race in their effect, and to the individual only as an integral part of the whole. To illustrate. The sign Aquarius is an electrical, positive, masculine influence, and will consequently manifest its chief activities upon the masculine qualities of the human soul; and to-day we have evidence of this in the gradual enfranchisement of woman, arousing the positive attributes of her nature in demanding equal rights with her brother, man, in the political arena, as she has already done in the educational field. The masculine portion of the race is becoming more aggressive, mentally, asserting greater individuality, independent thought and action. The intellect of the race is being directed, however slowly, into scientific channels, while the human soul is slowly awakening to a sense of a deathless immortality and a desire for spiritual truth. It is slowly but surely shaking off the yoke of an effete priesthood and the fetters of superstition and tyranny.

Intelligent man talks of the new scientific and intellectual era that has dawned upon the world; of the necessity for a new religious system, based upon scientific truths, which can be demonstrated, combined with the pure spiritual essence found in all systems of religion; a religion with more spirituality and less theology; a broader charity and less dogma, and deeper love for God and man, its only creed.

We must now consider the astral influence of the cycle upon the physical organism of mankind, and particularly of the Western races, who are moving upon the upward arc of the cycle. It is quite evident that a radical change must take place in the physical form and constitution with the influx of more intellectual, ethereal and spiritual vibrations. The organism must become more refined and compact, a greater degree of sensitiveness be attained, with a highly nervous system. The forerunner of this superior organism is now apparent in the numerous schools of physical culture and gymnasiums throughout the land, the many articles and pamphlets on deep, rythmic breathing disseminated among the people, and last, but not least, the various schools of mental healing, etc. The masses look on and wonder, while they exclaim: "What marvelous changes are coming to the world!" but are utterly ignorant of the cause of the mysterious change. To the student of Hermetic Philosophy there is no mystery involved. He knows the cause, and confidently watches for the effect.

Each one must seek to comprehend for himself, according to the light he may receive, basing his premises upon the TRUE PRINCIPLES of astrology, carefully noting the triune aspect of planetary influence upon humanity, ever remembering that the Sun and Moon are the great factors in human destiny, and that his premises and conclusions must occupy the same plane. Having acquired a knowledge of the science in its application to the individual, take the broader field, or universal aspect, as it applies to human races, and you will find the rise and fall of nations, empires and families marked upon the celestial dial, and in perfect accord with the influence of the Sun and planets upon Mother Earth, in her various movements. And last, but most important, seek with an earnest desire for truth to learn the relation of those glittering constellations of the shining Zodiac to the human soul and their influence in shaping its eternal destiny. This will reveal the whole of involution and evolution in a general sense.

A faithful, earnest and devout study of the "Science of the Starry Heavens" will lead us on to other planes of thought, relating to still more interior realms of knowledge than we perhaps now dream of, and, in the words of the master: "A true knowledge of the stars will include a true knowledge of the soul," and we shall realize "the mystical link that binds the soul to the stars."
MINNIE HIGGIN.

CHAPTER 1

THE ZODIAC

To the ordinary astrologer the Zodiac is simply a band of space, eighteen degrees wide, in the heavens, the center of which marks out the pathway of the Sun during the space of one year of 365 days, etc.

The twelve signs are to him simply thirty degrees of the space (12 times 30 equal 360), bearing the names of the constellations which once occupied them. Nay, he, as a rule, still imagines in some sense that the signs (constellations) are still there, and that the power and potency of the twelve signs is derived from the stars which occupy the Zodiacal band of the skies.

But this is not so, as any ordinary astronomer well knows. This single fact, i.e., the gradual shifting of the constellations, the DISPLACEMENT, let us say, of the starry influx from one sign to another without any ALLOWANCE being made in the astrologer's rules for any such change, has been one of the greatest obstructions to the popular spread of the art among EDUCATED MINDS. Argues the scientist: The "fiery influence of Aries," if depending upon the stars of that constellation, ought now to be shedding forth their caloric from the sign Pisces, and Aries ought to be lumbering along with the earthy Taurine nature. So, also, the lords of these signs ought to be changed, but that they are not can be proved by the fact that our earliest records of that dim, historic past show, equally as well as your latest "text-book," that Mars is the lord of Aries—a fiery planet in a fiery sign; but astrologers still say that Pisces is watery and Aries fiery, WHICH IS NOT THE CASE, IF THE STARS HAVE ANY INFLUENCE AT ALL. It is not necessary," say these logical thinkers, "to learn your abstruse science if we can demonstrate that the very basis upon which your conclusions rest is in every sense fundamentally false." The scientific facts of the case are as follows: The influence of the twelve signs, as described by astrologers, is a delusion, because in all ages they are reported the same; whereas WE KNOW that every 2,160 years each sign retrogrades to the extent of thirty degrees, and, as your art does not make allowance for this, it is false. For, if the influence of the twelve signs does not emanate from the stars occupying the space of those signs, it must emanate from nothing—a doctrine well suited, no doubt, to musty old sages of your superstitious Chaldea, but quite out of court in our progressive age—the last decade of our cultured and scientific nineteenth century.

So far, so good. And so the world rolls along its bright pathway in the heavens, little heeding the logical conclusions of an exact science. But to an initiate of those inner principles of our planet's constitution all these mental conflicts have a meaning and a purpose within Nature's divine economy; for it is neither wise nor expedient that the masses, with popular science in the lead, should grasp the truths which Mother Nature reserves ALONE for her own devoted priests.

The shining Zodiac, with its myriad constellations and its perfect galaxy of starry systems, derives its subtle influence, as impressed astrologically upon the human constitution, from the solar center of our solar system, NOT FROM THE STARS which occupy the twelve mansions of space. Aries, the fiery, and PISCES, the watery, ARE ALWAYS THERE, and, instead of its being an argument against astrology, it is one of its grandest truths that, in all ages and in all times, Aries, the first sign of the Zodiac has been found EVER THE SAME, equally as well as Pisces the last.

In order to convey our meaning, let us digress for a moment and bring forth a fitting illustration. The condition of our atmosphere and the surrounding objects—vegetation, etc.—have a peculiar condition and a magnetism wholly their own when surveyed exactly at sunrise. There is a freshness and peculiar sense of buoyancy not visible at any other time. If this state could be registered by any instrument and compared with any other set periods during the day, it would offer a remarkable contrast. Two hours later there is a very different influence, and at noon there is a wonderful contrast. The same may be said of sunset, and again at midnight; and, lastly, note the difference two hours before dawn. This is the coolest period of the whole twenty-four hours. These are facts, and yet our hearts are all beating to the same life-flow, and the Earth is no farther away from the parent Sun; and yet it is the angle at which we, THE INHABITANTS, receive this Sun's light that makes all the difference between dawn and sunset, noon and midnight.

When to these facts it is further added that it is sunrise, noon, sunset and midnight at the same instant, all the time, to some of the various, different portions of the globe, it demonstrates most conclusively that the Earth itself is enveloped, so to say, in a complete circle of conditions very similar to the twelve signs of the celestial Zodiac.

If we apply the foregoing illustration to the twelve signs of the Zodiac, we shall see a perfect analogy. We shall find that when the Sun reaches the celestial equator, so that it is equal day and equal night on the Earth, that he is on the line of the celestial horizon; it is cosmic sunrise. Hence Aries, the fiery Azoth, begins his active influx, and extends for thirty degrees, equal to two hours of the natural day.

It is the fiery red streams of awakening life that we all manifest at sunrise; then comes a change of magnetic polarity after the first fiery flush of cosmic life; the gleeful chattering of the birds and the cackling of the poultry. A reaction is noted; all things before active become restful and quiet.

So it is with vegetation, so it is with infant life, and so it is with cosmic conditions.

This corresponds with the sign Taurus. It is the solar influx, thirty degrees removed from his point of equilibrium toward the North. As this sign represents the powers of absorption, we see that at this period vegetable and animal life is quietly absorbing, for its own use, the fiery streams of solar life.

Again we view the activity of solar influx from a different angle and change of polarity, and all things become active, It is executive force. This corresponds to the sign Gemini. It is the solar influx, sixty degrees removed from his point of equilibrium. Then comes another change of magnetic polarity. It is rest from labor; it is noon. This corresponds to Cancer. The analogy is perfect. It is the solar influx, ninety degrees removed from his point of equilibrium toward the North, and the highest point in the arc of his apparent journey and of cosmic life. It is the equilibrium of life forces.

Again the fiery influx begins its activity, and, as the hottest part of the day is about two hours after noon, or middle of the day, so is solar influx most potent at this point in the Zodiac.

This corresponds to the sign Leo. It is the solar influx, removed 120 degrees from his point of equilibrium and thirty degrees toward the South. And so on month after month, until the last one, Pisces, which well corresponds to the watery skies of February and the lifeless period two hours before dawn of a new day upon the Earth, a new year to man and a new cycle in the starry heavens. The Zodiac, then, as it applies to the human constitution and the science of astrology, has its foundation in the Sun, the center and source of life to the planet; and the twelve signs are the twelve great spaces of our Earth's annual orbit about her solar parent, each one typical of its month, and each month typical of its corresponding action upon our Earthy conditions.

As each sunrise is different in its aspects, so are no two signs of the Zodiac alike. The sunrise on the first of March is wholly different from the sunrise upon the first of May. So is the beginning and ending of each sign, and the beginning and ending of each natural day, peculiar unto itself.

When we reflect upon the inner laws of this action and interaction, we come nearer and nearer to the one great occult fact, viz.: THE DIVINE ONENESS OF LIFE.

We find a perfect analogy between the destiny, the life, and expression of life on the Earth, and the life and material destiny of embodied man. He, too, has his sunrise, the beginning of a new day of life, the seedtime, the flowering season, when life wears a roseate hue; the ripening fruits of experience, his harvest-time—it may be tares or golden grain; his gradual decay, the ebbing of the life forces and the icy winter of death; his gentle zephyrs and destructive hurricanes, floods and tempests, periods of drought and plenty. Within his triune constitution there are spring tides and low tides of physical, intellectual and spiritual forces. Man also makes the annual journey about the solar center, when, at the beginning of each new year to him, the life forces of his soul are renewed, regalvanized, so to say, according to the magnetic polarity of his constitution.

And so, every form of life has its Zodiac, its orbit of life and destiny. It may be infinitesimal, or vast beyond conception, each in its own peculiar plane. So we see that, the whole visible universe is one vast organism, the medium of expression for the invisible, real universe—the soul and God, the great central Sun, the eternal center of all life, binding the whole into unity—ONE LIFE.

The celestial signs of the shining Zodiac have no existence to us apart from the graceful and unwearying motion of our Mother Earth. She alone makes our seasons, years and destiny; and she alone, by her motion about the Sun, determines the thrones and mansions of the planetary powers.

The astrological Zodiac of a Saturn or a Mars cannot be like ours. Their years and seasons are peculiar to themselves and their material conditions; hence the twelve constellations have no existence as objective facts of concrete formation or cosmic potentiality. No! But as unalterable symbols of occult truth, the starry pictures of the shining constellations have an eternal verity. They pertain to the living realities of the human soul and its varied experience.

What the mysteries are, and what connection they have with the twelve constellations, will form the subject of our next chapter.

CHAPTER II

THE CONSTELLATIONS

The twelve great constellations of the zodiacal belt which forms the Earth's orbit and the Sun's shining pathway around the celestial universe have been considered as mere imaginary figures, or emblems, invented by an early, primitive people to distinguish the monthly progress of the Sun and mark out, in a convenient manner, the twelve great divisions, or spaces, of the solar year. To this end, IT IS THOUGHT, the various star groups, termed constellations, were fancifully imagined to represent the various physical aspects of the month, under, or into, which they were consecrated by the Sun's passage during the annual journey, so that, in some sense, the, twelve signs or constellations were symbolical, not only of the seasons, but also of the labors of the year.

That such a system seems perfectly natural to the learned mythologist, and that granting the ancients so much is a very great concession toward this CHILDISH KNOWLEDGE is, of course, quite excusable when we are constantly told, or reminded, that actual science—that is to say, "EXACT SCIENCE," does not date backward more than a couple of centuries at most.

Even the modern astrologer, much as be descants upon the influence of the twelve signs, has but little, if any, real knowledge of this matter above and beyond the purely physical symbolism above mentioned. And perhaps it is as well that such a benighted condition prevails, and that the Divine, heavenly goddess is unsought and comparatively unknown. The celestial Urania, at least, in such isolation remains pure and undefiled. She is free from the desecrating influence of polluted minds.

Such, in brief outline, is the general conception of mankind regarding the shining constellations that bedeck, like fiery jewels, their Maker's crown, and illumine with their celestial splendor the wondrous canopy of our midnight skies. Is there no more than a symbol of rural work in the bright radiance of the starry Andromeda, the harbinger of gentle spring? Nothing, think you, but the fruit harvest and the vintage is in the fiery, flushing luster of Antares and the ominous Scorpion? Are men so spiritually blind that they can perceive nothing but the symbol of maturing vegetation and the long summer's day in the glorious splendor of Castor and his starry mate and brother, Pollux? It would, indeed, seem so, so dead is the heart and callous the spiritual understanding of our own benighted day. To the initiate of Urania's mysteries, however, these dead, symbolic pictures become endowed with life; these emblems of rural labor or rustic art transform themselves from the hard, chrysolitic shell and expand into the fully developed spiritual flowers of spiritual entities, revealing in their bright, radiating lines the awful mystery of the soul's genesis, its evolution and eternal progressive destiny amid the mighty, inconceivable creations yet to come; pointing out each step and cycle in the soul's involution from its differentiation as a pure spiritual entity, a ray of Divine intelligence, to the crystallization of its spiritual forces in the realms of matter and its evolution of progressive life; the same eternal symbols of the springtime, the glorious summer, the autumn and winter of its eternal being.

In making this attempt, probably the very first within the era, to convey in plain and undisguised terms the interior mysteries of the twelve constellations, the reader and student is advised to ponder deeply upon the outlines presented. The subject is too vast to present in one or two chapters. Therefore we hope that this revelation may incite the student to further research. The real significance, the true, spiritual importance of such mysteries, can only be realized and fully appreciated after prolonged meditation and careful study.

With this brief digression, which we consider needed advice, we will resume our task, and attempt to usher our student into the weird labyrinth of Solomon's starry temple—"the house not made with hands, eternal in the heavens."

1. Aries

"First Aries, in his golden wool."

This constellation represents the first Divine idea, the "word" of the Kabbalist, and the first active manifestation of the glorious En Soph. In other words, it is MIND IN ACTION, the first pulsation of Deity in the dual aspects of "Lord and Creator." To the human soul it is, and always typifies, the unknown, invisible power which we term INTELLIGENCE; THAT WHICH KNOWS, and gives unto each Deific atom of life that distinguishing, universal, yet deathless force which not only constitutes its spiritual identity and physical individuality, but enables it to pronounce, in the presence of its Creator, those mystic words: "I am that I am." In other words, this beautiful constellation symbolizes the first pulsation of that ray of pure intelligence which constitutes the Divine Ego of the human soul. It is the force that impels ever onward the life atom in its evolutionary progress, and reveals to us the beginning, or first manifestation, of the Divine Ego as an active, self-existing atom of Infinite spirit, within angelic spheres.

Seeing the actual, spiritual reality symbolized in Aries, how easy it is to note its full significance upon the external plane when refracted and reflected into the planes below through the complex action of the human organism, conveying the same radical influx in the first astrological month and the first sign of the Zodiac. We can read a perfect parallel in the astral influx upon the human body, as set forth in the "Light of Egypt," vol. I, which says Aries symbolizes the sacrifice and represents the springtime, the beginning of a new year. The first action of pure intelligence brought forth the first expression of form, and led to the sacrifice of its angelic state, and, having gained the victory over the lower realms of matter, once more the springtime of a new life, with the promise of life, light and love.

The sign Aries represents the thinking powers of humanity; in short, the active, intellectual being, the lord of material creation—Man; and in its cosmic relations, as shown under "The Occult Application of the Twelve Signs" (vol. I), we find the same perfect analogy.

II. Taurus

"He (Aries) turns and wonders at the mighty Bull (Taurus)."

The second constellation of the shining twelve represents the first reaction of spiritual conception. In other words, it is the mind's attention to its own ideas. In the Kabbalah it represents that peculiar state of executive force whereof it is symbolically said: "And the Lord saw that it was good," after each act of creation.

When intelligence first manifests itself form is a matter of necessity, and, as no form can possibly exist without matter, so Taurus is the first emanation of matter in its most etherealized state. Hence it is feminine, Venus the ruler thereof, and it represents the first pure form of the human soul, as it existed in its bright paradise within the angelic spheres of its parents, and reveals to us the first surprise of intelligence in embryo, the first sensation of consciousness, so to say—conscious of its Divine selfhood. Hence "He (the male spirit of pure fire, Aries), glorious in his golden (solar) wool, turns (expressing reaction) and wonders at the mighty bull (or material form)." Thus the first idea of pure intelligence in embryo, the result of action in Aries, becomes objective to its consciousness and is surprised at its own conception. It is the first sensation of pure, Divine love within angelic realms, and it (the male spirit of pure fire) sees that it is good.

Bringing this spiritual reality within our conception, and comparing it with its reflected astrological influx, what a beautiful harmony we find, and yet so simple that verily we cannot refrain from once more quoting our old-time, worn, yet, nevertheless, golden law: "AS IT IS ABOVE, SO IT IS BELOW; AS ON THE EARTH, SO IN THE SKY." Reflecting that Taurus is an Earthy sign, and a symbol of servitude, we see that matter is ever the servant of spirit, a necessary means for the manifestation of intelligence, again recognized in the fecundating forces of this astrological sign on every plane of its action. And it is ruled by Venus, the love element in Nature, her sympathies ever finding expression in this beautiful sign. What can be clearer, more understandable, than, that the involved principles and Deific attributes, as represented by the shining constellations, when refracted through the human organism, so complex in its constitution, reflects qualities which are the external and parallel expression of the subjective principles, and, further, that form is absolutely necessary for the manifestation of intelligence?

III. Gemini

"He (Taurus) bending lies with threatening bead, and calls the Twins (Gemini) to rise. They clasp for fear and mutually embrace."

This bright constellation (Castor and Pollux), Gemini, is spiritually representative of the second spiritual action. Hence it is, of course, a masculine sign and positive. We have witnessed act I of the soul's drama, and, as some have said, tragedy, and in this, the third of the shining twelve, we find the opening scene of act II, viz: The evolution of the twin souls, or, more correctly, the differentiation of the Divine soul into its two natural component parts—male and female.

Here we approach one of the most arcane secrets within the wide scope of Occult philosophy, hence must be exact, and at the same time clear, in our statements. Note, then, that after the male spirit of pure, ethereal, divine fire (Aries) bad conceived the first idea, and Taurus, the material envelope, had given that idea objective existence to its (the Ego's) consciousness, we find SENSATION AS THE RESULT. No sooner sensation than aspiration; i.e., longing. This closes the action and the reaction.

Ever, in obedience to the unsatisfied wants of an immortal soul Nature immediately responds. Hence "He bending lies with threatening head, (that is demanding)," and calls the twins (the twin souls) to rise (to appear or evolve forth)," and as a first rude shock caused by their separation, or, rather, by their separate existence as two distinct, yet mutually dependent, forces, we have the context.

"They clasp for fear and mutually embrace."

This most impressive scene in the soul's drama is one of profound interest and sublime beauty.

In the Kabbalah we find the same parallel, wherein it is stated: "And so God created man in His own image (the action of Aries and Taurus); in His own image (mind) created He him, male and female created He them." In other words, Aries, Taurus and Gemini are thus spoken of in pure allegory.

The mundane Bible of the Jews, like everything else esteemed sacred, finds its original and perfect expression in the great Astral Bible of the skies.

To the average student the evolution of the Twin Souls is a profound mystery, embracing, as it does, the whole of involution and evolution, seeing that this beautiful constellation represents to us the first recognition, or consciousness, of the Divine Ego of its dual forces, sensation and aspiration, called forth by the action of Aries and Taurus. How beautifully has the poet expressed this first pulsation of Divine love: "They clasp for fear," etc. Evolved by the Divine will of pure intelligence, they must ever remain as separate, yet mutually dependent, forces, positive and negative, male and female, upon whose action and reaction rest the perfect evolvement of the powers and possibilities of the One.

In order to clearly grasp the whole of these ramifications, we again invite our student's careful attention to the same sign, Gemini, in its astrological aspect, as it is representatively expressed by refraction upon the human organism. We find that this sign (the representative of the constellation always) signifies the union of reason with intuition, and that it governs the arms, hands and executive forces of man.

Surely, as we reflect upon the almost marvelous inter-relationship between things spiritual and things temporal, we must conclude, with the man Jesus, that "They have eyes but they see not, and ears, but alas they hear not."

If it were not so man would, indeed, by virtue of the latent forces within him, take the kingdom of Heaven by storm and reign supreme as enthroned king of all material forms. Man, in his blindness, has relegated intuition to obscurity; has neglected the cunning of the left hand and debauched the pure love of the divine state. Consequently, the executive forces within him are unbalanced, thus rendering him the slave of material forms, instead of being their lawful sovereign. Therefore, not until, with clean hands and pure heart, he restores intuition to her throne, united with reason, can he hope to COMPREHEND the reality of this arcane mystery of the twin souls, Gemini.

IV. Cancer

"And next the Twins with an unsteady pace
Bright Cancer rolls."

In this beautiful constellation we witness the reaction of Gemini, the closing scene in act II. Hence it is, of course, a feminine force we are observing. In other words, it is that period (or rather one of them) wherein the Kabbalah expresses the reaction of the En Soph, via his Creators, as "And behold the Lord saw everything that He had made, and behold it was very good."

Just so Cancer, spiritually interpreted, means equalizing, hence HARMONY, which is indeed very good as contradistinguished from chaos.

To the human soul Cancer is the period of exalted rest. It is the highest point in the arc of the Divine Soul's Angelic Cycle. From this glorious, but subjective, summit or altitude in the realm of spirit it must descend.

Restless energy and the still unsatisfied longings of its own immortal nature are the forces that bring such evolution about. Having evolved the dual forces of its divine nature, the Ego sees that it is good and rests from its labor. But as this exalted state is purely subjective, and ideal, it must of necessity, to satisfy the longing for further unfoldment and desire to know, descend into material realms and conditions. From this point begins the soul's involution downward, until the lowest point in the arc is reached, viz., Capricorn.

Refer now to the sign Cancer, and carefully study out the parallel upon its astrological planes and also under its Occult aspects, as given forth in the "Light of Egypt," Vol. I, where we read: "Cancer rules the respiratory and digestive functions of humanity, and governs the reflective organs of the brain." Note the parallel. Within subjective realms the Divine soul has inspired and assimilated all that is possible to that angelic state, and knows a period of blissful rest. But the longings of its immortal nature urge on the soul. So we see that the sign Cancer symbolizes tenacity to life; to live we must breathe and eat and assimilate upon every plane of our being. It necessarily follows that, the mentality expressed by Cancer must be susceptible to inspirational currents; to inspire is to indraw. In its application, we find that this sign symbolizes love. How beautiful the harmony and contrast of the constellation and its astrological representative.

V. Leo

"Then Leo shakes his mane."

Herein is typified the third grand spiritual action which, as we find throughout Nature, travel in pairs; hence Leo is a positive, masculine constellation.

Having attained the highest point in the super-celestial states of subjective, embryonic existence; having evolved sensation and aspiration; now, inspired by a desire for immortality, the DUAL SOUL of the Divine

Ego is once more impelled forward; but, as all evolution works in spirals, it cannot ascend higher without first apparently descending lower; so ever onward in its eternal march. This beautiful constellation symbolizes the first action on the downward portion of the are. It is the affinities of the heart, so to say, working from within to without.

Matter, in its more etherealized form, begins to assert its sway. The allegorical serpent of Eden is working upon the feminine portion, symbolized by the heart, and, like a magnetic tractor, the soul's affinities are drawn downward, and, as if in defiance of all responsibilities, consequences, and Karma, the soul, lion-like, "shakes his mane in the imperiousness of deathless courage."

As we read these weird allegories, written by Deity in the starry vaults of heaven, the interested soul bows in reverence and awe before that almighty power we term Providence, and the profane call God.

No man has altered these pure records of divinity; no finger has interpolated one single line. They are as beautifully clear to the soul now as they were in the very dawn of Nature's awful creation.

To the Initiate into Urania's mysteries it is unnecessary to draw a parallel between the constellation and its astrological sign. They are too clear, magnificent and impressive to escape notice. To the majority of students the resembance may not be so apparent, hence, for their benefit, we will point out a few aspects of this interesting parallel.

We read that the sign Leo is the "solar Lion of the mysteries, that, ripens with his own internal heat the fruits brought forth from the Earth by the moisture of Isis (the soul)." Just so, the Divine Ego, by its eternal energy and strength, the pure fire of intelligence, externalizes through material forms the principles involved in the downward portion of the arc, as qualities and attributes of the soul (reflected in the physical man as traits and qualities). Again we are told, "this sign reveals to us the ancient sacrifice and the laws of its compensation." In the imperiousness of a deathless courage, the soul defies all consequences and responsibilities. Surely, this is the supreme sacrifice, to leave its pure, Edenic state to gain knowledge, to evolve its latent forces. And from this lion of the Tribe of Judah, is born that Divine love and sympathy which ultimately redeems and purifies the soul and saves it from death in matter. The laws of its compensation are fulfilled in the prefected man.

In its intellectual aspect, we learn that the mental forces of those dominated by this sign are ever striving to attain unto some higher state. Their ideas are grand, compared with the nature of the constellation, and all that it implies. The reflection is clear, natural and beautiful. When we reflect upon this awful period in the involution of the dual souls of the Divine Ego, as symbolized by the constellation, and the grand truths represented by the astrological sign when refracted through the human organism, the reason for Leo being named the Royal Sign becomes quite plain.

VI. Virgo

"And following Virgo calms his rage again."

Beautifully expressive are these lines to those who read their mystic import aright. Virgo is the reaction of the leonine force, and is, consequently, a feminine symbol.

Action and reaction are the eternal laws upon which the cosmos is founded. They constitute the inseparable affinities, attraction and repulsion, of everything within the realm of manifested being. In this mystic constellation, we see the first ideas of maternal instinct arise. This is a necessary result of the impulsive action of the heart in Leo—the reaction from a state of imperious, defiance. The heat of rage or energy and deathless courage results in the IDEAS of something to be encountered, overcome, and of self-preservation. The dual soul descends still another volve in the spiral of its celestial journey toward crystallized forms.

Virgo, the Virgin of the skies, and eternal symbol of that Divine, immaculate conception, shows wherein these forces lie. Here is conceived, in a pure, holy sense, the first instinct of love within the dual soul. It represents that awful period in the Biblical Garden of Eden, wherein the VIRGIN WIFE stands before the tree of knowledge, of good and evil, where she is fascinated by the allurements of matter and is unconsciously becoming enveloped in the coils of the serpent. In other words, after the cosmic force had SHAKEN ITS MANE in defiance of material forces, it is the reaction of his subjective half which sees HOW GOOD material things are; or, in other words, "and following Virgo calms his rage again." The masculine half, or positive force of the soul, yields to temptation and is soothed by the alluring prospects.

It will be noticed in this connection that pity, reflection, and compassion, are the peculiar actions of the sign Virgo in the Zodiac (not the constellation), and that astrologically it governs the bowels. This symbolism is really very beautiful when closely compared and studied. That immaculate conception of pure love of the soul for its other half, upon the astrological plane, becomes refracted and reacts as compassion and pity. Again, the soul, within subjective realms, sees how good material things are, and its refraction represents the assimilating functions of the human organism. It also reveals to us the significance of the Lord's Supper. At this stage of its journey, the Divine Ego knows for the last time that close communion with the twin soul before the crucifixion, the wine typical of the sacrifice, the bread, and the sustaining forces, of its own immortal being.

The intellectual aspect of the sign Virgo forms a perfect analogy to the constellation, and is too evident for further remarks.

VII. Libra

"Then day and night are weighed in Libra's scales;
Equal awhile, at last the night prevails."

Another volve in the spiral, and we reach the grand climax of the soul's journey, within the spiritual world.

The nature of this constellation was, for ages, concealed from all but Initiates; for the reason that, it contains the most important mysteries connected with the human soul. It is the grand transition are between the spiritual world and the astral world; in other words, between ideal conceptions and elemental forms, between the world of design and the realms of force.

One of the chief mysteries of Libra is, that, it is androgyne, or bisexual, in nature. So far the dual soul has evolved within the realms of spirituality; here it stands, in the celestial balance, between the two, giving way to temptation, takes the forbidden fruit and instantly awakes from its purely spiritual state to become surrounded by the illusions of matter. The struggle of the soul with the attracting forces of matter is very clearly expressed in the line:—

"Equal awhile, at last the night prevails."

In other words, astral and physical darkness bedim the soul's spiritual sight, and, leaving the realms of innocence and bliss, they sink into the vortex of the great astral world.

The celestial state is now forever lost as a realm of angelic innocence. It can only be regained amid trial, sorrow, suffering, and experience, and, when regained, it is as Lord and Master, not as the innocent cherub. But when, having gained or reached the equator of the upward arc of its progress, and, united once more to its missing half, gives expression to that deathless force with which it started from the opposite point, Aries: "I AM THAT I AM;" no longer an embryo, but being within the UNIVERSAL SOUL of being. Before closing this symbolic constellation, we must reveal the mystery of its BI-SEXUAL NATURE. In the higher or first portion of the sign it is {}, positive to some extent, and masculine. The soul is still within the Garden of Eden and pure, clad in the raiment of God, and is represented by the Chaldean statues of "The Bearded Venus," or Venus, the Angel of Libra, as a morning star, bright Lucifer. But in the latter half, after the fruit of the tree of knowledge of good and evil (positive and negative, you see) has been partaken of, bright Lucifer falls. The Sun of the Morning, shorn of his glory, becomes the symbol of night, or Vesper, the evening star, and the symbol is thus {}, and the soul loses its heavenly raiment, or spiritual consciousness, and becomes clothed with matter, the symbol of night.

The sign Libra in the Zodiac, in its astrological aspect, is a very external correspondence of all the foregoing.

VIII. Scorpio

"And, longer grown, the heavier scale inclines,
And draws bright Scorpio from the winter signs."

We now behold the gates of Paradise guarded by the FLAMING SWORD which points to the four quarters of the world. This sword is, according to Genesis, "to guard the way of the tree of life," and such, esoterically, it really is.

The soul is no longer dual, but separated into male and female personalities; "and behold they see that they are naked." Stripped of their spiritual raiment, they feel the chill of matter and the lusts of an animal nature. They need clothing, "so God made them coats of skin." Sex is the symbolism herein typified, and the evolution of the animal passions of procreation, of multiplication and evolution. It is the complete entry of the soul into elemental conditions, and the flaming sword guarding the four quarters of the Earth to the way of life are the four great realms of the astral world; the way to physical life in concrete forms; and the way to life eternal through the realms of the Sylphs, Gnomes, Undines and Salamanders. They are the basis of all matter, known as Air, Earth, Water and Fire. Here we see that, through the evolution of sex and its accompanying desire for procreation, these blind forces of Nature find their avenue of expression. Spiritual consciousness almost lost, and without reason, the soul becomes the prey, so to say, of these forces of the astral world, which is the realm of design. The soul's creations must be met and vanquished upon the upward arc of the Cycle of Progress. They guard or oppose the way to eternal life.

Here the soul, having gained the victory, stripped forever of its earthy raiment and the lusts of the flesh, arrayed once more in its spiritual raiment, purified and sanctified, it will stand once more at the gates of Paradise, where, reunited with its missing half, it will partake of the fruit of the Tree of Life and become as Gods. Astrologically the correspondence is perfect, and so thinly disguised as to need no explanation.

IX. Sagittarius

"Him Centaur follows with an aiming eye, His bow full-drawn, and ready to let fly."

Deeper and deeper sinks the soul into material forms. The evolution of sex has produced the necessary avenues for the entrance of countless forces, and the soul is now rapidly losing the last vestiges of its spiritual conscience. In other words, Sagittarius symbolizes that state of the soul wherein it is descending to its polarizing point, and is, therefore, the vortex of innumerable opposing forces, seeking expression in different forms.

"It is the bow (strength or force of the soul), FULL-drawn and ready to let fly" its arrows (of energy) in any direction that may afford proper opportunity. Here we see the expression of that deathless, fiery force, and imperious daring and courage, within more material states; the primal fire reflected from another angle.

But everything is unsettled. It is a masculine force, and restless, and is represented under the allegory of the "Tower of Babel" and the utter dispersion of the people (entities) to the four corners of the Earth, and finally becomes involved in dense matter, and its migrations are at an end on this side of the Cycle of Necessity.

Upon the astrological plane, the zodiacal sign Sagittarius rules the motive forces and the pedestrian instinct, the thighs, or basis of locomotion; hence, we see, even here, a most perfect analogy. This sign symbolizes, also, the governing forces of humanity, which see the necessity of law and order; hence government. In this expression, we find the bow (strength or force) ready to let fly its arrows of energy in any direction the opportunity may afford; when refracted upon the human organism and reflected upon the external plane, these forces manifest as the restless spirit, that ever impels onward, seeking new fields of expression, out of which develops a sense of order, restraining and training, or the governing of self and control of others. When we reflect upon these symbols of starry truths the mind bows in reverence before the wisdom that created them.

X. Capricorn

"Next narrow horns the twisted Caper shows."

The Goat, and in the realms of spirit, the crystallized mineral is the reaction of the former, and shows to us death, inertia and rest; hence Kronos, or Saturn, the symbol of death, is lord of this state and condition. It is the polarizing point of the soul's evolution in matter, and therefore, forms the lowest are in the Cycle of Necessity.

Herein we behold the soul, imprisoned within the mineral state. The fire of the flint, and the spark in the crystal, are the only avenues of its lonesome expressions. But, as the lowest point, it is also the promise of a higher, and the symbol of a higher state, and the symbol of another spiral in its endless life.

This constellation, as the symbol of inertia and death, is also the symbol of awakening life, and prepares the soul for the more perfect expression of its powers in its forthcoming upward journey. If we pause for a moment and consider the force and power necessary to evolve out of this dark, dense, mineral realm, the foregoing sentence will become clear and forcible. Hitherto, the soul has been slowly drawn down into coils of matter, imprisoned by material forces. It has penetrated the lowest depths, and can go no farther. Rest here, is to gather strength, force. Mark well the difference and parallel between Cancer and Capricorn, opposite points in the arc. Cancer is the symbol of exalted rest within angelic realms; Capricorn the symbol of rest in dense matter. From the former state the soul is impelled forward on its downward journey; from the latter state the soul awakens to the struggle for life on the upward are; and must now give expression to the positive powers of its immortal being, which have become involved in material form; that shall make it the master, and give it the victory over death and material forces. Surely, this is truly the promise of a new day, and higher state of existence.

It will he instructive to study this by a comparison of the zodiacal sign, Capricorn, as set forth in the "Light of Egypt," Vol. 1, wherein we read: "This sign signifies the knees, and represents the first principle in the trinity of locomotion, viz., the joints, bending, pliable, movable." The analogy is perfect. The soul, which has been pliant, bending to material forces, now reverses this action, and bows the knee in awe and reverence to the higher powers of its being. When refracted upon the human organism, we find that the cold, lonesome state, and weary struggle within the mineral realm, becomes love of self, directing its energies to the attainment of selfish ends. What could be more natural?

XI. Aquarius

"And from Aquarius' urn a flood o'erflows."

The soul, released from its crystallized cycle of matter, now rapidly evolves into states, though material, yet entirely different. Its previous arc, from Libra to Capricorn, has been amid inorganic matter. It is now rushing with lightning speed upon its weird, toilsome, upward, journey through purely organic forms, from vegetable to animal; and, as all organic forms have their primary origin in water, so does this celestial urn express the primary conception of this physical state. Further, to more fully express this, Aquarius is typical of man, as prototype of the last grand goal of the soul's future material state—in other words, the last quadrant of the four elements, viz.: Bull, Lion, Eagle, Man.

There is something exceedingly significant in all this, and the more we ponder on this spiritual allegory of the shining constellation, the more we are impressed with the divine wisdom of those early instructors of our race, who thus preserved truth in an incorruptible form.

From this weird, but beautiful constellation, we learn how the soul has progressed, finding innumerable avenues of expression of its latent forces; the manifestation of its powers in the various chemical changes, and development of functions expressed through countless forms, on the lower planes of existence. The sacrifice of its angelic innocence, the imperious defiance and deathless courage, symbolized by Leo, have obtained the victory over the lower kingdoms; which will be incorporated into his vast empire. Yet, unstable as water, it cannot excel; or, in other words, cannot rise to a higher state within this are, of its progressive life.

We find that the astrological expression of this constellation, the sign Aquarius, governs the legs, and is the natural emblem of the changeable, moveable, migratory forces, of the body, forming a perfect parallel with its interior symbol. There is a great deal contained in this zodiacal sign worthy of deep study and reflection.

XII. Pisces

"Near their loved waves cold Pisces take their seat,
With Aries join, and make the round complete."

Once more a reaction—the last scene of the soul's impersonal drama. The constellation (if Pisces is the symbol of rest and expectation. The soul has now completed the first round, or rung, in the Cycle of Necessity; and its next state is that of incarnated man. It has triumphed over every sphere below, and defied, in turn, every power above, and is now within that sixth state of the embryonic soul-world that transforms all its past knowledge, sorrow, and suffering, into experience; and produces the impersonal man.

It has traveled through constellated states within matter and spirit, and, as a human soul, with reason, intuition, and responsibility, it will, in its next state, become subject to those same powers when reflected from a different plane. The twelve constellations of its soul will manifest a complete rapport with the twelve signs of solar light and power.

With this we close. The mystic sign of this constellation is {}, or completion, a seal and a sign of its past labors.

And, as we have seen, the shining constellations are the soul's progressive history from its genesis, to its appearance within embodied conditions as man; and so, by correspondence, are the twelve solar signs symbols of man and his material destiny. The foundation has been laid, the material and resources are at hand, for his kingdom is exclusive. With his own hands he must build his temple (the symbol of the perfected man), each stone accurately measured, cut, polished, and in its proper place, the proportions symmetrical, hence, harmonious; the keystone of whose arch is WILL, its foundation love. This accomplished, be will have completed the second round of the great Cycle of Necessity.

And who, after contemplating the wondrous harmony of this beautiful system, and the complete accord of each part, can refuse to agree with the truly inspired Addison that—

> "Ever moving as they shine,
> The hand that made us is divine."

CHAPTER III. THE SPIRITUAL INTERPRETATION OF THE TWELVE HOUSES OF THE HOROSCOPE

As a sequel to the foregoing subjects, viz., the Zodiac and constellations, we will add the spiritual interpretation of the twelve houses of an horoscope, which completes the triune expression of these celestial symbols of eternal truths.

In revealing this mystery, we would impress upon the mind of the student that the order of the Zodiac is the reverse of the external, in its spiritual application, to the twelve houses of the horoscope.

As the four cardinal signs, viz., Aries, Cancer, Libra and Capricorn, correspond to the four angles of a natal figure, it is our purpose to explain, first, the symbology of the four angles, or cardinal points; believing the whole revelation will thus become clear and forcible.

The four angles of the horoscope correspond to the four elements, the four triplicities, and the four cardinal points, or epochs, in the soul's involution from pure spirit to the crystallizing, inert, mineral state.

The first angle is the ascendant, or House of Life. It is the eastern horizon, and symbolized by Aries. Upon the interior, this first angle stands for the birth, or differentiation, of the Divine Ego, as the result of the creative action, or impulse, of the Deific mind.

The Ego rises upon the eastern horizon of celestial states, a glowing, scintillating atom of pure intelligence, an absolute, eternal Ego, rising out of the ocean of Infinite Love.

The South angle, meridian, or Tenth House, pertaining to honor, etc., is symbolized by Cancer; the highest point in the arc of the soul's involution, as a differentiated atom of Deity within angelic spheres.

Having evolved the first dual expression of its (the Ego's) self, the twin souls—Sensation and Aspiration, or Love and Wisdom, the Ego rests awhile, radiant with celestial love and wisdom, and inspiring the Divine breath of life.

Again the restless impulse of the creative purpose arouses the Ego to further action. The culminating point has been reached, and now must begin an apparent downward course toward the western horizon.

The seventh angle, or House of Marriage, etc., is represented by Libra (the Balance), or point of equilibrium; where the two souls are still one, balanced upon the western horizon. The alluring temptations of material illusions draw the souls downward, and, divorced from their celestial state, the radiance of Divine love becomes obscured, until the twilight of consciousness of that former state is lost in the night of material conditions.

This house signifies, also, law, and open enemies, and (Libra) justice. Sex is the law. The antagonism is surely too apparent to require explanation.

The fourth angle, or Nadir, the point opposite the M. C., signifies the frozen North, and is symbolized by Capricorn, the crystallizing point in the soul's involution. It is death, inertia; that is, crystallization of the soul's spiritual forces. It is the lowest point of the are in the monad's downward journey. It is the night, before the awakening of a new day upon a higher plane of existence.

The remaining houses are the lights and shadows that, fill out and complete the picture, upon this, the first round of the Cyclic Ladder.

The Twelfth House, symbolized by Taurus, represents the first expression of form of the human soul. It is matter in the most etherealized state. It is the trail of the serpent; the silent, secret, tenacious, negative principle; that ultimately draws the soul down into the vortices of gross matter and death.

The Eleventh, or House of Friends, whose symbol is Gemini, the Twins, expressive of the first emanation of this sublime relationship, the dual attributes, love and wisdom, closest friends. It is sensation and aspiration, which enable the spirit to attain to the exalted state indicated by the Tenth Mansion.

The Ninth Mansion of the celestial map is the House of Science, Art, Religion, Philosophy, etc., and its symbol is Leo, the Heart, with its emotions, love, and longings, and sympathies. Having evolved the twins, and inspiring the Divine breath of wisdom; glowing with Deific love, the Ego aspires to know; and all the sympathies of the soul are aroused. Dauntless and fearless, defying all opposition and consequences, It (the Ego) is ready to sacrifice this angelic state and explore the boundless Universe in pursuit of knowledge, and goes forth on its long voyage upon the ocean of Infinite, fathomless love and wisdom.

The Eighth, or House of Death and Legacies, is symbolized by Virgo, the virgin wife, standing before the "Tree of Knowledge of Good and Evil," fascinated by the flattering prospects of greater power and wisdom. Desire and sympathy draw the soul down into realms which lead to death, and the beginning of a heritage of sorrow,

The Sixth, or House of Sickness, Menials, and Sorrow, is symbolized by Scorpio. The fall, from Libra through Scorpio (sex), created the first condition of what we recognize as sickness and affliction. It is evident that this house is related to the elementals of the astral plane, which become the servitors of man.

The Fifth, or House of Children, etc., symbolized by Sagittarius, signifies the offspring of sex (Scorpio), entities sent forth to people the Earth, to take their chances of life, speculating on its future course, fearlessly eager for the struggle, gaining pleasure in its migrations and activities.

The Third House is symbolized by Aquarius. This is the first step of the upward journey, or evolution, from the inert mineral state. The changes are now rapid; the journeys innumerable; through mineral, vegetable, and animal planes, of existence. Here, the Soul Monad brings into actual practice the knowledge gained on its long voyage. The magical powers of the soul are brought into action to effect these changes in form and function, conquering material forces and planes of life, transmuting Nature's elements to its uses and purposes, and writing its history, as it journeys ever onward, step by step.

And further, this house stands in opposition to the Ninth House, symbolized by Leo; longing to expand its (the

Ego's) possibilities through trial and suffering; gaining knowledge through bitter experience; yet fearlessly braving all things; guided and sustained by the imperial will of spirit. The recompense promised by that supreme sacrifice has been won in Aquarius—the Man—consecrated now to a higher existence, baptized in the waters of affliction (experience), ready to be transmuted into actual knowledge. This is Aquarius, and the Third House.

The Second House, signified as Pisces, the House of Wealth, that which has been accumulated on the long and toilsome journey—the wealth of experience, acquired through trials and struggles. And now, with higher, greater possibilities, the soul eagerly awaits the hour when it shall be born again, a conscious, responsible human being, to begin the second round of the Cyclic Ladder; on this second round, to externalize the knowledge gained, to evolve the involved attributes and forces of being,—a creature of will and intellect, to work out its destiny, as the lord of material creation.

Observe, the order of the Zodiac is reversed upon the external human plane. But, Aries is always symbolical of the first angle, and Libra of the seventh, being the point of equilibrium, while the tenth, or South, angle becomes Capricorn and the fourth Cancer. The mission of the soul now is to evolve the positive, spiritual attributes.

Aries rules the brain and the fiery, imperial will. It signifies courage, daring, etc., the first qualities necessary for the battle of life. Ruling the head, the sign and house show us the ability of man to view the field of action, to mark his chart, and arm for the war (which will be incessant); responsible for his acts, a creature of unfolding consciousness, an individual, whose measure of free will enables him to wander so far North or South of his celestial equator, within his orbit, or Zodiac.

The South angle, or Tenth House, now ruled by Capricorn, tells of the honor, position, fame, etc. (or the reverse) acquired by patient labor. The crystallized material gains, the concrete result of ambition, skill, and talent, which will, at the close of his earthly career, become liquefied by the universal resolvent; symbolized by Cancer upon the opposite angle, symbolical of the grave, the end of mundane affairs; when they will be mirrored forth in new forms in that great white sea, according to the manner in which he gained his worldly accumulations and prestige.

The seventh angle is Libra, House of Marriage; that all-important relation which may make or mar a life, the Balance is so easily disturbed in its equilibrium. To preserve its harmony, equality must reign, blending love and wisdom. It is the perfect poise of body, mind, and soul, achieved by loving obedience to the higher laws of our being and the true union of intuition and reason.

The Second House, now represented by Taurus, shows us that personal wealth and possessions must come through patient servitude, steady application, and diligence, in being able to choose and assimilate the knowledge, that will enable man to battle with material conditions, and wrest from the abundant sources of Mother Nature his share of treasure and experience. It is the battle-ground to which humanity, armed with brain and will, life and energy, goes forth to battle with material forces for the bread he must earn by the sweat of his brow, and through the silent, subtle forces of mind and soul conquer matter, thus storing up a wealth of knowledge and experience.

The Third House is ruled by Gemini, the Twins, Reason and Intuition, the brethren who aid and guide us on our many journeys in the pursuit of knowledge. As this sign governs the hands and arms and the executive forces of humanity, we see, that, the hands become the magical agents of mind, moulding into outward form the ideas conceived in the mind, projecting these into the field of active life, that he may write a bright record in the Book of Life. The hands should be kept clean, the images pure; and the perfect poise gained by the equal exercise of love and wisdom, intuition and reason, making the basis of education; the evolution of the interior or real self. This is the true meaning of this house upon the external plane. it is Occult, because it means projecting the powers of the soul into conscious life, externalizing the qualities and magical forces of spirit, as shown in the first instance by Aquarius. This can be accomplished only through pure desire and aspiration. Otherwise, the unbalanced scales, with floods and cataclysms, will be the results.

The Fifth House, ruling children, etc., is symbolized by Leo (the Heart). The joys and sorrows that offspring of every kind bring, all belong to this House of the Heart. The sacrifice indicated is too obvious for comment.

The Sixth House governs sickness, disease, etc., and its symbol is Virgo, an Earthy sign, clearly showing us that the material form is the matrix, out of which are born disease and suffering. But, the perfect assimilation of the fruit of the "Tree of Knowledge of Good and Evil;" transmuting the trials, experiences, sorrows, and suffering of the physical and external life into true wisdom; makes man master of his material universe; and the blind forces of Nature become his servants. Having accomplished the task, and attained the harmonious poise, or balance, in Libra, the individualized soul arrives at the eighth step in the journey.

The Eighth House, or House of Death and Legacies (Scorpio). The old Adam dies. The sensuous has no place in the balanced, harmonious being, but recognizes sex as the law, the door to regeneration now, and that a new legacy is awaiting him.

The Ninth House, Sagittarius. where, with the knowledge acquired, self sits in judgment upon the works of the hands and mind, whether or no they have been well done; the sacrifice of the lower nature properly made, control of the triune being established, the transmutations correct and accurate, assimilation perfect and free from dross, harmony gained by loving obedience to the higher law of being, and thus becomes ruler of his kingdom; the long journey almost accomplished, so far as Earth is concerned, perceiving and understanding that all sciences, philosophies, and religions, have their origin from one primal source. Having penetrated the depths in reverent obedience to the Divine law of creation, and evolved the attributes of the dual constitution, the forces of his being become crystallized. He has reached the culmination of his earthly pilgrimage, and stands forth the perfected human reflection of the higher self.

The Eleventh House is represented by Aquarius on the human plane. Friends surround and welcome him. These friends are the pure thoughts, noble impulses, lofty ideals, and generous deeds. The bread cast upon the Waters of Life returns to nourish and sustain him in his encounter with the secret foes, symbolized by the Twelfth House and Pisces. The idols, false ideas, and vampires of his own creation, are to be cleansed and washed away by the Waters of Love, the universal solvent that is ever seeking to bring about change and new forms; born again of water to make the round of the astral Zodiac, until, having again reached the equator of the ascending are, where he is reunited to the missing half of his soul, the true friend of the Edenic state; the highest point in the arc of human progress won; the honor and glory of a perfected soul; the Lord and Master, the "I am that I am," to rest in peace in the heart of Infinite Love and Wisdom.

CHAPTER IV. ASTRO-THEOLOGY

There is one species of Divine revelation which has not, and cannot, be tampered with, one great Bible, which forms the starry original of all Bibles.

This sacred Bible is the great Astral Bible of the skies; its chapters are the twelve great signs, its pages are the innumerable glittering constellations of the heavenly vault, and its characters are the personified ideals of the radiant Sun, the silvery moon, and the shining planets, of our solar sphere.

There are three different aspects of this sacred book, and in each aspect the same characters appear, but in different roles, their dress and natural surroundings being suited to the natural play of their symbolical parts. In fact, the whole imagery may be likened unto a play, or, rather, a series of plays, performed by the same company of artists. It may be a comedy, or it may be melodrama, or it may be a tragedy; but the principles behind the scenes are ever the same, and show forth the same Divine Oneness of Nature; demonstrating the eternal axiom. ONE TRUTH, ONE LIFE, ONE PRINCIPLE, AND ONE WORD, and in their fourfold expression, is the four great chapters of the celestial book of the starry heavens.

In this aspect the visible cosmos may be represented as a kaleidoscope. The visible constellations, planets, and other heavenly bodies, are the bits of colored glass; and Deity the invisible force, which keeps the instrument in motion. Each revolution produces a different pictorial figure, which, complete in its harmony of parts, is perfect in its mathematical proportions, and beautiful in its geometrical designs. And yet each creation, each form, and each combination of forms, are produced by the same little pieces of glass; and all of them, in reality, are optical illusions; i.e., natural phenomena, which deceive the physical senses. So it is with Cosmic Nature.

It must not, however, be supposed, because of this perfect and continual illusion of Nature's playful phenomena, that all visible creation is purely an illusion of the senses, as some cranky metaphysicians would have it, because this is not so.

Going back again to our kaleidoscope, we can clearly see that without it, and its tinted beads, no such optical illusion is possible. There is, then, a basis of spiritual reality to all visible physical phenomena; but this basis lies concealed, because of the perfect illusion which the reflected image produces upon the material plane of the physical senses. The beads themselves are real. These are the basis, and the different pictures are the result, not of the beads, but of the angle from which they are reflected to our earthly vision. In other words, THE PLANE FROM WHICH WE BEHOLD THE PHENOMENA.

Hence, the nearer we approach the Divine center of our being, the less complicated Nature's original designs become, and the farther we are removed from that central source, the more weird, mysterious, complicated, and incomprehensible, does Mother Nature appear, to the finite human mind. And this is especially so, to man's theological instinct, his religiosity, that constitutes one of the fundamental factors of his being.

Nature is ever one in her original truths and their duplicate reflections; but ever conflicting and contradictory in her multiplied refractions through the minds of men. Therefore, we will present the primary concept of that grand Astro-Theology formulated by man's great progenitors; and view the simple machinery, by which they typified to the primitive mind a general outline of Nature's Divine providence.

All sacred books begin with an account of physical creation, the culmination of which, is the appearance of man and woman, as the parents of the race; and, while they will differ considerably in detail and make-up, the basic ideas embodied are essentially the same in all cosmo-genesis; so that in the Jewish Bible, accessible to all, one can read the primitive story of creation from a Jewish point of view, and, when read, rest satisfied that he has read the revelation vouchsafed to man in every age and in every clime. The only difference is one of mental peculiarity and national custom, along with climatic conditions. Hindoo, Chaldean, Chinese, Persian, Egyptian, Scandinavian, Druidic and ancient Mexican are all the same—different names and drapery, to suit the people only, but essentially the same in the fundamental ideas conveyed.

THE CREATION OF THE WORLD

The simple story of creation begins at midnight, when the Sun has reached the lowest point in the arc—Capricorn. All Nature then is in a state of coma in the Northern Hemisphere, it is winter time, solar light and heat are at their lowest ebb; and the various appearances of motion, etc., are the Sun's passage from Capricorn to Pisces, 60 degrees, and from Pisces to Aries, 30 degrees, making 90 degrees, or one quadrant of the circle. Then begin in real earnest the creative powers, it is spring time. The six days are the six signs of the northern arc, beginning with the disruptive fires of Aries. Then, in their order, Taurus, Gemini, Cancer, Leo, Virgo; then Libra, the seventh day and the seventh sign, whose first point is opposite Aries and is the opposite point of the sphere, the point of equilibrium, equal day and equal night, it is autumn. It is the sixth sign from Aries, the first creative action, and so the sixth day following the fiery force, wherein God created the bi-sexual man. See Genesis, 1:5-27: "So God created man in His own image; in the image of God created He him, male and female created He them."

It is the seventh, or day of the Lord (man), the climax of material creation and Lord of all living things, and be rests in the blissful Garden of Eden. This seventh day and seventh sign is the concealed sacred Libra the perfect union of the sexes. Then comes the fall from Libra, through Scorpio, and banishment from the Garden of Eden. That is the victory of Satan, or Winter, over Summer, etc. It is useless to repeat the same old, old story. The yearly journey of the Sun around the constellated dial of Deity is the Astro basis of all primitive cosmology.

THE SCHEME OF REDEMPTION

In addition to the creation of the world and the fall of man through sin, we find all people in possession of a grand scheme of redemption, and, like the former, we shall find them all essentially the same. They all require a mediator between the angry God and disobedient man, and they all require that this mediator shall be Divine, or semi-Divine. Nothing less can satisfy Deity's demands; or, rather, let us say man's own carnal imagination. It is simply another turn of our cosmic kaleidoscope, and behold! the actors have changed. Capricorn becomes the stable of the Goat, in the manger of which the young Savior of the world is born. As a type of all, we will take the Gospel Savior. It is again midnight. The Sun enters the sign Capricorn on the twenty- first of December. This is the lowest point of the arc, South, and for three days he is stationary, or in darkness. And now it is Christmas Eve. He (the Sun or Savior) begins to move, and at midnight is born as the celestial Virgo is rising upon the Eastern quadrant of the skies; hence the Sun-God is born of a Virgin. Then comes the flight to escape Kronos, or Saturn (ruling Capricorn), who kills the young babes. There is a period of silence in the God's history while the Sun is in transit through the signs Capricorn and part of Aquarius. That is, he is hidden or obscured by the clouded skies of this period. We hear of him but once again until he, the Sun-God, or Savior, is thirty years old, or has transited thirty degrees of space. He has entered the sign Aquarius (symbolical of the Man.) Now begins the period of miracles.

Let us digress for a space, and refer to our chapter on the constellations. We shall find a perfect analogy between this miracle- working period and the constellations Aquarius and Pisces, as therein given. The first miracle we read of is turning water into wine. This may be seen in a threefold aspect. The Sun-God changes by his life-forces the waters of winter into the rich vintage of the harvest, where the Virgin (Virgo) Mother again appears. Again, the wine becomes the blood—the life offered up on the vernal cross to strengthen, renew and make merry with new life our Earth and its people. The devil (or winter), with his powers of darkness, is defeated and man saved. The final triumph is the crucifixion in Aries, the vernal equinox, about the twenty-first of March, quickly followed by the resurrection, or renewal of life. Then the God rises into heaven, to sit upon the throne at the summer solstice, to bless his people. We read, that, the Savior of mankind was crucified between two thieves. Very good. The equinoctial point is the dividing line between light and darkness, winter and summer. In other words, the Sun is resuming his northern arc, to replenish the Earth with his solar force and preserve his people from death in the coming winter. The life of a Buddha, a Krishna or a Christ, are all found in their completeness in the life of Horus; while the Father, Son and Holy Ghost are Isis, Horus and Osirus. The same trinity, under different names, are found in all nations. It is the Sun, Moon and Human Soul, which is the only true mediator of Man.

There is another version of this celestial crucifixion, wherein the Sun-God-Savior, after the supper of the harvest in Virgo, is crucified at the autumnal equinox upon the equator. We read that he was dying from the sixth to the ninth hours—three hours, three signs, or from the 21st of September to the 21st of December, when he is laid in the tomb. This is the lowest point of the Sun's journey in the southern hemisphere, and darkness holds the balance in our northern hemisphere. The three days in the tomb are the three months, or three signs, before the vernal equinox, or the resurrection, the rising out of the South to bring salvation to the northern portion of our Earth.

We have now only to glance over various diverging lines of the same cosmology and the same redemption. All these allegories typified TRUTHS. They all teach the Initiate the mysteries of creation, of man's destiny and his necessary Cycle of Material Probation. Some of the most beautiful parables may be read in this light. Abraham, and the story of his wanderings in the deserts of Asia Minor; of Lot and his unfaithful wife, are to be seen still written in the heavens. Hagar and Ishmael are still there; so also are Esau and his brother Jacob; the story of Joseph and his brethren; of Sampson and his twelve labors. This is the same beautiful story. The Sun, shorn of his glory, or solar force, at the autumnal equinox, stands upon the equator between the two pillars of the temple (or light and darkness), and pulls down the temple (or signs) into the southern hemisphere. And behind this we have the eternal truth of the soul, when, giving way to the allurements of matter (Delilah), the soul is shorn of its spiritual covering, or conscience, and sinks into matter and death. And the story of David and Goliath can be read to-day as clearly as of yore.

They are eternal, spiritual verities of human nature, and record, not only the history of the human race, its mutations and transmutations, but of the individual man and the suffering and delusive joys of his material life. Aye, more! It is the record of all his past existence and a type of his eternal destiny in the future.

Another turn of our cosmic kaleidoscope, and lo! the scene changes —the play extended, the angles greater, caused by the revolution of our solar parent through his celestial Zodiac. As the Sun passes out of one sign into another, or, in other words, forms a different angle to his own center of force, a new dispensation is born to the world; or, rather, re-born under a new guise. The great Sun-God appears to change his nature and manifests an entirely different set of attributes. That is the way man personified this play of Nature, through his imperfect conception of the cause of this change. But to him it was, and is, a truth, and man's effort to externalize these attributes in a Divine personality was, and is, strictly from the plane of his mental development and spiritual unfoldment.

The two pictures of this Astro-Theology, as set forth in the two divisions of the Jewish Bible, will illustrate our meaning. The Sun had entered the sign Aries some time prior to the exodus from Egypt. Aries is the constellation of Mars, the fiery, destructive and warrior element, or force, in Nature, and we find the Jewish conception of God a perfect embodiment of these attributes: The Lord of Hosts, a God mighty in battle, delighting in the shedding of blood and the smell of burnt offerings, ever marshalling the people to battle and destroying their foes and the works of his own hands; a God imbued with jealousy, anger, and revenge. This was the type set up by the Jewish savior and lawgiver, Moses.

After a period of 2,160 years, we find the Christian cosmology ushered in. The Sun has entered the sign Pisces, which is ruled by Jupiter, the beneficent father. The Christ, or mediator, of the Christian Gospel was an embodiment of the joint qualities of the sign and ruling planet. Gentle, loving and merciful, His words were messages of love and peace; His work was with the poor, oppressed and fallen; he eschewed sacrifices and burnt offerings; a contrite heart was the best offering; He taught the people that God was their Father, loving all, just, yet merciful. But a strong taint of the old conception has remained with the human race, hiding, at times, the beauty of the latter concept. These are, again, the refractions of eternal truths, viewed by man from his material plane. The elements are here presented, the alphabet and its key clearly defined. Therefore, let each one explore this tangled labyrinth of Astro-Theology for him or herself, and work out the various correspondencies at leisure. It is enough to indicate the starry originals of all this seemingly confused mass of so-called Divine revelation in sacred books.

They, one and all, pertain to the same celestial phenomena, and the various Bibles are the outcome of man's serious attempt to tabulate and externalize this heavenly order, to record his conceptions of these starry aspects and movements with their corresponding effects upon the Earth.

Probably the purest system to us is that which may yet be derived from Chaldean sources. This sacerdotal caste were the most perfect in their astral conceptions and complete in their symbolic system of recording, and if the great work found in King Sargon's library in seventy tablets is ever translated, it will prove of priceless value to the student of these weird, but sublime, astrological mysteries.

In conclusion, as we reflect upon the fourfold aspect of the subject that we have presented in outline in these pages, the whole imagery passes in review before the mental vision. We see that the radiant constellations of the heavenly vault, with the beautiful reflection and counterpart, the shining Zodiac, are the two halves of the great Cycle of Necessity, the spiral of eternal, universal life, which binds the whole into unity, and unity into infinity. It is the grand scheme of creative life. The seven principles of Nature, or Divine Activities, are the forces producing the phenomena within seven angelic states, seven kingdoms, and, by seven planets, upon the external plane; the planets being the passive mediums of the positive spiritual forces. Upon this dual spiral, which reflects the seven rays of the solar spectrum is produced seven musical notes; one half of the spiral in sound and color being the complementary of the other half. Man, the Earth, and our solar system, are revolving, each orb in its own key, and its own peculiar ray, meeting and blending with other spirals, and the whole blending into one mighty spiral Cycle of Progressive Life, revolving around the Eternal, Infinite Ego-God, ever involving and evolving the attributes, powers and possibilities of the One great central source of Being.

It is a grand orchestra, pealing out in richest melody and sublime HARMONY, the grand Anthem of Creation: "We Praise Thee, O God."

CHAPTER V. ASTRO-MYTHOLOGY

The Astro-Mythological system of the ancients, though forming the last section, so to say, of the mysteries of the Divine Urania, is, perhaps, the most beautiful of its general features, and perfect in the complete fulfillment of the purpose for which it was intended, viz.:— to convey to the human mind a lesson, a moral, a truth in Nature; and last, but not the least, to serve as a basis upon which its inner aspirations and its more external faith might rest in security.

When we come to examine the deep, philosophical principles of such a wise system, we are almost astounded at the result of our researches and the wisdom of human nature displayed in formulating such perfect analogies of truth, semi-truth, and of falsehood, according to the plane occupied by the individual.

Let us take one instance, which will clearly explain all the rest, for they are built and formulated after the same model. Aeneas, of Greek myths and fables, is reputed to be the son of Venus by a MORTAL father, upon the plane of reality. As that of actual PARENT and CHILD, of course this is an utter falsehood. To the rural population of long, long ago, and their simple, rustic conceptions, IT WAS A TRUTH.

Why so? Because they believed it, and to them it taught the required lesson of obedience to the powers that be. But if in reality it was a falsehood, how can it become a truth by the simple addition of acceptance and belief? Because it possessed a metaphysical truth, though not a physical one, in the sense accepted.

Aeneas, son of Venus, whose history is so beautifully preserved by the immortal Virgil, was (metaphysically speaking) son of the goddess, because he was, in his astral and magnetic nature, ruled and governed by Venus, born under one of her celestial signs and when she was rising upon the ascendant of the House of Life, even as Jesus Christos was born of a virgin, because Virgo was rising at His birth.

Thus Aeneas was, in strict metaphysical reality, a son of Venus. Having satisfied the rural mind, which thus, unconsciously, accepts an absolute truth under a physical disguise, the metaphysical thinker, the philosopher, also accepts the same fable, knowing and realizing its more abstract truth,

But, again, we are met with the objection that such a truth is only apparently a truth; i.e., on the plane of embodied appearances, and naturally the question arises, where (if at all) is the real truth of the mythos? That truth which is beyond the mere metaphysical thinker and commonplace philosopher; the truth which the Initiates recognize—where is it? That truth lies far beyond the purview of Astro-Mythology. It is connected with the center of angelic life. Sufficient here to say that, as there are seven races of humanity, seven divisions to the human constitution, seven active principles in Nature, typified by the seven rays of the solar spectrum, so are there seven centers of angelic life, corresponding to the seven planetary forces formulated in "The Science of the Stars," and, as each one of us must of necessity belong to one of the particular angelic centers from which we originally emanated, the Initiate can see no reason why AENEAS MAY NOT IN REALITY belong to that celestial vortex represented by Venus upon the plane of material life. This being the case, we see how beautiful the ancients' system of temple worship must have been. The simple rustic, in reverence and awe, accepting the gross and physical meaning—the only one possible to his dark, sensuous mind. The scholar and philosopher bow their wiser heads with equal humility, accepting with equally sincere faith the more abstract form of the allegory; while on the other hand, the priest and the Initiate, lifting their loftier souls above the earth and its formulas of illusion and matter, accept that higher and more spiritual application, which renders them equally as sincere and devout as their less enlightened worshipers. It is thus we find these astro-myths true for all time, true in every age of the world, and EQUALLY TRUE OF ALL NATIONS. And this is the real reason why we find every nation under the Sun possessing clear traditions relating to the same identical fables, under different names, which are simply questions of nationality. And when mythologists, archaeologists, and philologists once recognize the one central, cardinal truth, they will cease to wonder why nations, so widely separated by time and space, possess the same basic mythology. They will then no longer attempt its explanation by impossible migrations of races, carrying the rudiments with them. They will find that this mythology was a complete science with the ancient sages, a UNIVERSAL MYSTERY LANGUAGE, in which all could converse, and that it descended from the Golden Age, when there was but ONE nation on the face of the Earth, the descendants of which constituted the basic nucleus of every race which has since had an existence. In this light all is simple, clear, and easy to comprehend—all is natural.

The astronomer-priests of the hoary past, when language was figurative, and often pictorial, had recourse to a system of symbols to express abstract truths and ideas. In order to impress the minds of pupils with a true concept of the attributes of the celestial forces, we call planets, they personified their powers, qualities, and attributes. Just as the average mind of to-day cannot conceive of

Deity apart from personality, so did primitive man clothe his ideas in actual forms, and in these impersonations, they combined the nature of the celestial orb with that of the zodiacal sign or signs, in which the planet exerted its chief and most potent activities. For instance, the planet Mars, whose chief constellation is Aries, was described as a great warrior, mighty in battle, fierce in anger, fearless, reckless, and destructive; while the mechanical and constructive qualities were personified as Vulcan, who forged the thunderbolts of Jove, built palaces for the gods, and made many useful and beautiful articles. Then, again, we find that Pallas Athene was the goddess of war and wisdom. She sprang from the head of Zeus. Aries rules the head, and represents intelligence. Athene overcame her brother Mars in war, which shows that intelligence is superior to brute force and reckless courage.

Here, we see three different personages employed to express the nature of the powers and phenomena produced. They were called gods and goddesses. This was quite natural, as the planets of our system are reflections of Divine principles. Esoterically, Mars symbolizes strength, victory—attributes of Deity.

Mars is said to have married Venus, teaching us that the union of skill and beauty are essential in all artistic work.

Mythology tells us that the god Mars was supposed to be the father of Romulus, the reputed founder of Rome. Romulus displayed many characteristics of the planet. The mythos is no doubt a parallel to that of Aeneas.

Rome was founded when the Sun in his orbit had entered the sign Aries, and Mars was the god most honored by the Romans. In time, with the degeneration of human races and their worship, to the rural mind, the subjects of the mythos became actual personalities, endowed with every human passion and godlike attribute, the former characterizing the discordant influence of the heavenly bodies upon man.

Gai, Rhea and Ceres, or Demeter (Greek), represent the triune attributes of Mother Earth. Gai signifies the Earth as a whole, Rhea the productive powers of the Earth, and Ceres utilizes and distributes the productive forces of Rhea.

In the charming story of Eros (Divine Love), son of Mars and Venus, he (Eros), we are told, brings harmony out of chaos. Here, we see the action of Aries and Taurus, ruled respectively by Mars and Venus.

The beautiful myth of Aphrodite, born of the sea foam, is Venus rising out of the waters of winter, to shine resplendent in the western skies at evening, and typifies the birth of forms, as all organic forms have their origin in water.

In all lands the Sun was known under various names, typical of solar energy, especially in reference to the equinoctial and solstitial colures.

Henry Melville, in his valuable work, "Veritas," says no reliance can be placed upon ancient dates, either of

Europe, Asia, or anywhere else, and he conclusively shows that such dates are Astro-Masonic points on the celestial planisphere, the events recorded being, as it were, terrestrial reflections of the celestial symbols.

To attempt to wade through all the various systems of mythology, and explain each in its proper order, would be to write a large encyclopedia upon the subject. We have given a few examples as keys, and suggest works for study. We have here given the real key, and the student must fathom particulars for himself. The chief work, and most valuable in its line, is Ovid's "Metamorphoses." The next, also the most valuable in its line, is "The Mythological Astronomy of the Ancients," with notes (these latter are the gist and constitute the real value), by S. A. Mackey; and last, and, perhaps, in some sense, not the least, is the "Wisdom of the Ancients," by Lord Bacon. This is published in "Bacon's Essays."

A careful study of Ovid, with the key which this chapter supplies, will reveal ALL that pertains to ancient gods, demi-gods, and heroes, while a study of Mackey, and a careful comparison with "La Clef" and "La Clef Hermetique" will reveal all that pertains to cosmic cycles and astral chronology, which is the only chronology that is quite trustworthy, as far as ancient history is concerned.

While we are on this subject, we must point out some of the delusions, into which the subtle, magical teachings of the Orient would lead the student.

All the monster sphinx, half human, half animal, etc., which the ancients have preserved, are simply records of the past. They are chronological tables of cosmic time, and relate to eras of the past, of the Sun's motion, and not by any means to living creatures of antediluvian creations, as some wiseacres have imagined. Many of these ancient monuments, monstrous in form, are records of that awful period of floods and devastation known as the Iron Age, when there was a vertical Sun at the poles; or, in other words, when the pole of the Earth was ninety degrees removed from the pole of the ecliptic. To those who can read aright, every lineament tells as plainly as the written word the history of that awful past, marking the march of time, recording the revolutions of the Sun in his orbit of 25,920 years, and relating with wonderful accuracy the climatic changes, in their latitudes, which took place with each revolution of the Sun and corresponding motion of the Earth's pole of less than four degrees. All the greater myths of the dim past were formulated to express cosmic time, solar and polar motion, and the phenomena resulting therefrom. These monuments of antiquity prove that, the ancients knew a great deal more of the movements of heavenly bodies and of our planet than modern astronomers credit them with.

Madame Blavatsky, in her "Secret Doctrine," seriously states that all these monstrous forms are the types of actual, once living physical embodiments, and, with apparent sincerity, asserts that the Adepts teach such insane superstitions.

Such, however, is not the case, neither is there anything true, or even approaching the truth, in the cosmogony given in the work in question.

And, lastly, we have but one more aspect of the grand old Astro- Mythos to present to your notice. This aspect reveals the whole of the ancient classification of WORK and LABOR, and gives us a clear insight into the original designs, or pictorial representations, of the twelve signs and the twelve months of the year. It also clearly explains many things which are to-day attributed to superstitious paganism.

As each month possesses its own peculiar season, so are, or were, the various labors of the husbandman, and those of pastoral pursuits, altered and diverted. Each month, then, bad a symbol which denoted the physical characteristics of climate and the temporal characteristics of work. As the Sun entered the sign, so the temple rites varied in honor of the labors performed, and the symbol thus became the object of outward veneration and worship. So we see that the twelve signs, and principally the four cardinal ones, became Deities, and the symbols sacred, but in reality, it was the same Sun to which homage was paid.

There is a large sphere of study in this direction, as, of course, each climate varied the symbol to suit its requirements. In Egypt there were three months when the land was overflowed with water; hence, they had only nine working months out of doors, and from this fact sprang the Nine Muses, while the Three Sirens represented the three months of inactivity in work, or three months of pleasure and festivity.

Mackey tells us that the great leviathan mentioned in the Book of job was the river Nile.

In nearly all mythologies, we find that the gods assembled on some high mountain to take counsel. The Olympus of the Greeks and Mount Zion of the Hebrew Bible mean the same, the Pole-Star; and there, on the pictured planisphere, sits Cephus, the mighty Jove, with one foot on the Pole-Star and all the gods gathered below him. The Pole-Star is the symbol of the highest heaven.

With this we close, leaving the endless ramifications of this deeply interesting subject to the student's leisure and personal research, trusting the keys we have given in this chapter and their careful study may induce the reader and student of these pages to search out for himself the meaning concealed in all Astro-Mythologies.

CHAPTER VI. SYMBOLISM

At this point of our study it is necessary to make a halt; and, before proceeding further, to attempt to formulate and realize that, which, so far, we have been pursuing.

First, then, we have passed in review the Zodiac, and then the constellations. From this we mentally surveyed both Astro-Theology and Astro-Mythology; and now, it is our first duty to realize these in their real significance, and this consists in a clear comprehension of the Grand Law of Correspondences.

What is this law? It is the law of symbolism, and symbolism, rightly understood, is the one Divine language of Mother Nature, a language wherein all can read, a language that defies the united efforts of both time and space to obliterate it, for symbolism will be the language of Nature as long as spirit expresses itself to the Divine soul of man.

No matter where we turn nor where we look, there is spread out to our view a vast panorama of symbolic forms for us to read. In whatever form, angle, or color they present themselves, the true student of Nature can interpret and understand their symbolic language aright. It has been by the personification of Nature's symbols, that man has become ignorant of their language. There is no form, sound, nor color but what has its laws of expression; and only a perfect knowledge of symbolism will enable man to know the law, power, and meaning, lying behind such manifestations. The law of expression is exact, and as unalterable as Deity Himself. The physical senses cannot vibrate to these interior forces, and through them, comprehend their law. The physical senses vibrate to the spirit's expression, not to the powers, forces, and laws, which brought them into objective existence.

Countless numbers of mystics, if such they deserve to be called, among present-day students, speak and write very learnedly upon the "Law of Correspondence," and few, if any, of them really understand or know anything at all of that law. The intellect alone cannot solve the problems of this law. It cannot grasp the true, interior and spiritual meaning, except in just so far as intellect is capable of externalizing them. The inmost spiritual truths, that cannot be demonstrated to the outward senses, never have, nor never will, appeal to any one who has not the interior ability to comprehend them.

There was a time when men ruled by pure intellect, without its accompanying other half, intuition: they were looked upon as monstrosities. This state of purely intellectual development has been brought about by the positive, masculine principle, reason, absorbing its counterpart, the intuition, the feminine portion; and the result, by correspondence, is as fatal as upon the interior plane, where the positive, masculine soul denies the existence of his mate; thus setting upon his throne, only a portion of himself as his idol, and then, reasons himself into the belief that he is complete. Love has been cast out, ignored and forgotten until at last she departs, leaving a vacancy, that eternity cannot fill.

This is somewhat similar to their illusive Devachan, an ideal, a mere mystical sentiment to gush over, but a something they do not in reality comprehend. Therefore, we shall do our utmost to explain this universal law, and to point out wherein its first principles are manifest. Once these are mastered, the Golden Rule will explain all the rest: "As it is below, so it is above; as on the earth, so in the sky."

Here, then, is our first lesson on the subject of REALITY, which constitutes the Hermetic science of Correspondences.

First, realize that a line or an angle, for instance, is something more than its mere mathematical outline. It corresponds to some power, force, or principle within the great Anima-Mundi of the mysteries, that are trying to find expression, in their evolutionary journey, in forms. Let us illustrate our meaning. A point or dot is what? Well, externally it is the alpha of all mathematics. It is the first finite manifestation of the spiritual force. Within that dot lies concealed, in embryo, all the future possibilities of the manifesting principle.

This dot or point is a something to begin with, a form externalized, from which all future forms may spring forth, and they may be infinite, both in number and variety. First a primary, simple idea, from which all ideas and thoughts, intricate and complex, have their being.

A point extended is a straight line, scientifically expressed (whereas in real truth there is no such thing as a straight line); that is to say, it is a form increased or multiplied by itself, and therefore, is an extension in space that can be measured, and each extension means a new form, an additional symbol. It has taken on new aspects, new relations, hence contains the second principle of mathematics, so to say; but, besides being points, THEY ARE SYMBOLS. They are principles in

Nature as clearly related to each other as the leaf and the stem of plant life.

Each monad, or point in the universe, is the beginning of something; equally so, it is also the termination of its own forces in that particular action, and will remain inert until it becomes acted upon by something else.

A point, then, is a primary, simple idea, a straight line. An angle is the same idea, rendered greater and more complex, and refers to the same forces upon a different plane, and the more we multiply the angles the more complex and far-reaching becomes the symbol and the more numerous and diverse become its planes of action. Here we will introduce an example. A trine represents three forces or angles, and, when united, form a trinity, hence harmony. Its apex (when above) is celestial, therefore represents the male forces of spirit.

A trine reversed also represents the same forces, with its apex in matter, hence it is negative. In these two complex ideas, clearly represented by these symbols, we have ALL matter and spirit; and yet they are but extensions of our point in space, rendered far-reaching and complex, by the position and the number of angles presented.

Let us turn the key once again, and we find that, both spirit and matter possess the same outline in their primal concept, except reversed (polarized).

Let us unite these two trines, and we have a still more potent form; a symbol almost infinitely complex. We have spirit and matter united, or, rather, three rays of force, positive, meeting three rays of force, negative, at a given point. Thus we have six points, also six sides, the ultimate of which is a cube. All are now equal. It is the first force of a crystallization (creation) of matter.

Once again let us turn the key, and we have our two conceptions in a metaphysical sense; the trine with its apex above is as the trine with its apex below, both the same in form, yet vibrating to very different planes, and a very different language is required to read and interpret their meaning aright. The spiritual, or the trine with its apex above, draws its influence from the celestial, and as it condenses and takes on form in the trine of matter, it transmits this same Divine force through its apex, which points below, to matter. The double trine is found upon every plane, obeying the Divine Law of Correspondences.

It is, in this sense, called "Solomon's Seal," because it is the grand hieroglyphic of the Hermetic law: "As it is above, so it is below; as on the Earth, so in the sky."

To continue this line of reasoning, or speculation, let us say, would lead us beyond the firm basis of human reason; it would escape the grasp of intellect, to which I am compelling this course of instruction to bend, but it would never take us beyond the real limits of the universe; yet, not to extend our investigations, we would ever remain in the lower trine, in the realms of effects, and lose sight entirely of the trine of spirit, from whence originated the force and potency in the form of matter.

Therefore it is, that the science of Symbolism has been evolved and formulated. The symbols, the manifestations, are ever present, and the study of effects will, to a developed soul, suggest the cause, the nature of the principle back of it, as well as the law which would produce such effects. Such is the science of symbolism; and it bounds and binds back into a religio-philosophical system, each class of symbols and each plane of manifestation; as securely as modern savants have defined the province of chemistry, magnetism, and mathematics; and, so far, these bounds are useful. But it is only a question of time and space, after all, because, when resolved back, or, let us say, evolved up into their abstract principles; chemistry, magnetism, and mathematics, are purely arbitrary terms to express special features of the same one eternal thing or science—which is EXISTENCE; and it, in turn, is CONSCIOUSNESS; not that external consciousness of existence, not that knowledge and love of living, but that interior, conscious knowledge, which tells us why and how we exist, by what force and power we are sustained and permitted to obey and carry out the law of mediumship, reception and transmission, attraction and repulsion, spiritual and material, that ultimately blend and become as one, the double trine, and, united with the Divine Ego of its being, becomes complete; seven, the perfected number of form.

The sum total, then, of all, and the value it may possess to the individual, is measured by his ability to perceive; for there is nothing external that is not in some sense mental, and there is nothing mental that is not in some sense spiritual. The sides of the triangle, physical, mental, and spiritual, and the apex where meet the mental and spiritual, forms the center of contact to higher trines in realms above.

Where mind is not, there are no symbols, no ideas, no manifestations. The spirit has not yet reached that point in its evolutionary journey where it can yet crystallize its projected force, power, or ideas, into forms; for everything that is, is the outcome of Divine thought, and expresses within itself the symbol of its being. This is the arcana of the Law of Correspondences.

Remember the above teaching, because upon its full comprehension rests the ability to read symbols aright. It will aid the soul to fully realize that, the vast universe is but the mental image of the Creator; that there is no such thing as manifested existence apart from mind; and consequently, the infinite worlds that float securely in space, blushing and scintillating with light of life and love of the Father, revealing to mortal minds some faint conception of the awful resources and recesses within Nature's star-making laboratory, are but the scintillating reflection of life, the reactions of mental phenomena. So, too, with the mental creative powers of the mind of man, for, not a vibration that proceeds from his every thought but what creates its correspondence in the creative realm of spirit. Hence, symbolism continues giving to the soul of man, throughout eternity, food for thought and contemplation.

All symbols, then, are objectified ideas, whether human or Divine; and as such possess a real meaning; and this meaning is altered, extended and rendered more complex with every additional thing or influence by which we find it surrounded, or with which, we find it correlated. For instance, $1.00 means one dollar; add six ciphers to the left 0000001., and it is still the same $1.00, and no more, because their position is previous, or before, the 1. But add the same number of ciphers to the right, $1,000000, and lo! we find a wondrous change of force, power, and consequence. We see all the mighty power of our million of money, and the possibilities and responsibilities with which, in these days, it becomes associated.

So it is with everything else in Nature. Man pays the penalty by increased responsibility, for every step in knowledge that be takes, as well as every dollar in gold be procures. Dollars, as well as talents, have to be accounted for, and their usefulness increased tenfold. The dollars must not be buried nor hoarded any more than our talents, but each, unfolded and doubled, so that we may be instrumental in helping our coworkers in their upward path, in the Cycle of Necessity. Knowledge is the basic foundation in reading Nature's language. Purity of thought, truth in motive, and unselfish benevolence, will lift the veil that now lies between the two trines, cause and effect, spirit and matter.

We have given the key and explained the alphabet of this wondrous law; therefore we close. Each must, by the same rules, work out the special links in the chain for him or herself. The angle from which each take their view determines the reading and interpretation of the symbols presented, whether that be from the apex, the sides or the base, for every symbol has its trinity in principles and form. Cause and effect are but the action and reaction; the result is the symbol which reveals the correspondence of both.

CHAPTER VII. ALCHEMY—PART I

What a weird yet strangely pleasing name the term Alchemy is. It is simple, yet so infilled and intermixed with the possible verities of exact science and the philosophical speculations on the infinite and the unknown, as to elude our mental grasp, as it were, by its own subtle essence, and defy the keenest analysis of our profoundest generalizers in science. And yet, in spite of this self-evident truth, how fascinating the sound of the word becomes to the mystic student's ear, and bow pregnant with awful and mysterious possibilities it becomes, to the immortal powers embodied within the complex human organism termed man.

Words, if we but knew it, have the same innate, magnetic influence, and possess the same power of affinity and antipathy, that the human family possesses; as well as all organic and inorganic forms and substances; and how sad, to a developed soul, to witness the inharmony existing in our midst, caused by the misapplication of names.

Most human beings are very conscious of personal, or human magnetism, and its effects. But they stop right there, and do not dream of the subtle, silent influences emanating from a name, a word, and the power existing in words, when properly used. The human mind is so absorbed in Nature's manifestations, which are only the husks, that they fail to see the true, hidden meaning and realities, concealed beneath the material shell.

We will first notice the meaning of the words which constitute our subject, viz., Alchemy, then give a brief review of its physical correspondence, chemistry, and its true relation to its spiritual counterpart, Alchemy.

"Al" and "Chemy" are Arabic-Egyptian words which have much more in them than appears upon the surface, and possess a far different meaning from the one which the terms usually convey to the average mind. Terms, and the ideas we associate with them, vary according to the age in which we live. So with those, from which the word Alchemy is derived.

Let us penetrate beneath the mere verbal husk with which linguistic usage and convenience have clothed them, and which, in the course of ages, has become nothing but the dross of decomposed verbiage, and see if we can excavate the living germ, that has become buried within. If we can do so, we shall, at the commencement of our study, have attained unto a realization of the ancient meaning and real significance of the terms employed. And this will be no small gain, and will form no unimportant part of the equipment in our present research.

The Arabians, who derived the whole of their Occult arcana from the Egyptians, are the most likely to render us the most truthful and direct significance of the word, and so we find them. Thus, "Al," meaning "the," and Kimia," which means the hidden, or secret, ergo THE OCCULT, from which are derived our modern term Alchemy, more properly Al Kimia. This is very different from the popular conception to-day, which supposes that the word relates to the art of artificially making gold by some chemical process, and viewing it only as some sort of magical chemistry, forgetting that, the science of chemistry itself is also derived from the Kimia of Arabian mystics, and was considered as one and the same thing by every writer of the Middle Ages.

At this time, the physical man was not so dense and grasping for husks; hence the soul and spiritual part had greater control, and could impart the real, the alchemical side, of Nature to him; hence the Law of Correspondences was understood, and guided the educated in their considerations, researches, and conclusions.

Do you ask why, if they were so enlightened, they have veiled their knowledge from the world at large?

The power of mind over matter was as potent in those days as now, and the masses were as correspondingly corrupt as they are today. Therefore, to put this knowledge into the hands of the multitude would have been generally disastrous. So they wrote it in mystical language, knowing that all educated students in Nature's laws, at that time, would understand; yet they little dreamed how much their language would be misunderstood in the centuries to follow, by those who look to their ancient ancestry for aid on subjects that have become at the present day so lost in mystery.

Having ascertained, beyond question, that Alchemy was, and consequently is, the secret science of Occultism—not the philosophy, mind you, but the science; let us proceed, for, we shall find that these two aspects may often differ, or appear to differ, widely from each other, though they can never do so in reality, for the latter produces and establishes the facts, while the former occupies itself in their tabulation and deductions. The science constitutes the foundation, and the philosophy, the metaphysical speculations, which rest thereon. If these important distinctions are borne in mind, all the apparent confusion, contradiction, and other intellectual debris, will either disappear or resolve themselves into their own proper groups, so that we may easily classify them.

It is at this very point, that, so many students go astray amid the labyrinths of science and philosophy. They, unconsciously, so mix and intermingle the two terms, that nine-tenths of the students present only one side of the question—philosophy, which soon runs into theory, if not supported by the science, which they have lost in their volumes of philosophy.

You may say, one subject at a time. Yes, this may be true, if its twin brother is not absorbed and forgotten.

In this chapter, we shall deal especially with organic Alchemy.

Organic Alchemy deals exclusively with living, organic things, and in this connection differs from the Alchemy of inorganic matter. These two aspects may, in this one respect, be compared to organic and inorganic chemistry, to which originally they belonged; as astrology did to astronomy. Alchemy and astrology—twin sisters—were the parents of the modern offspring, known in chemistry and astronomy as exact science. These latter, however, deal with shadows and phenomenal illusions, while the former concern the living realities, which produce them. Therefore, there can be "no new thing under the sun," saith Solomon.

First, let us deal with the most lovely form of our art, that which pertains to the floral and vegetable kingdoms. Every flower or blade of grass, every tree of the forest and stagnant weed of the swamp, is the outcome of, and ever surrounded by, its corresponding degree of spiritual life. There is not a single atom but what is the external expression of some separate, living force, within the spaces of Aeth, acting in unison with the dominant power corresponding with the type of life.

If science could only behold this wonderful laboratory within the vital storehouse of Nature, she would no longer vainly seek for THE ORIGIN OF LIFE, nor wonder, what may have become of the missing link in scientific evolution, because, she would quickly realize that, biogenesis is the one grand truth of both animate and inanimate Nature, the central, living source of which is God. Science would also, further realize that, this biune life is ever in motion throughout the manifested universe; circulating around the focii of creative activities, which we term suns, stars, and planets, awaiting the conditions which are ever present for material incarnation; and under all possible combinations of circumstances and conditions, conceivable and inconceivable, adapting itself to continuous phenomenal expression. Links, so called, in this mighty chain of evolution, may appear to be missing here and there, and, for that matter, whole types may seem to be wanting, but, this is only because of our imperfect perception, and, in any case, can make no real difference with the facts, because, if such be a reality, if there be what we may term MISSING LINKS in the scheme of evolution, it only shows that spirit, although associated with, is ever independent of matter.

But matter—what is to become of it? Is it independent of spirit? The kindness of the Divine spirit heeds not the unconscious mind of matter and its boasted independence, and works silently on, and at last, accomplishes its mission—the evolution of matter, the uplifting of the soul of man, as well as the universe. The blindness of man is dense, and the saddest part to admit is that, they will so stubbornly remain so.

If, for one instant, the penetrating eye of the soul could shine forth through the physical orbs of vision, and imprint the scenes, beheld behind the veil, upon the tablets of the brain of the physical organism, a fire would be kindled that, could never be quenched by the fascinating allurements of the material, perishable things, of matter.

That development of the real atom of biune life can, and does, go forward, irrespective of the gradation of physical types, needs no convincing proof, other than visible Nature.

MAN IS NOT THE OUTCOME OF PHYSICAL EVOLUTION, and produced by a series of blind laws, that lead him upward from protozoa to man, as a child climbs up stairs, advancing regularly, ONE STEP AS A TIME. This latter conception, we know, is the theory of exact science, but not of Alchemy, not of the science of Occultism. Man, according to Wallace, Darwin, Huxley, and Tyndall, is what progressive stages of physical evolution have made him. But the very reverse is true. The fauna and flora of past geological periods are what the human soul has produced, by virtue of its gradual advancement to higher states and conditions of life, so that, so far from man being the outcome of the planet's development, such material progress is the outgrowth of man's advancement, proving again that, matter is not independent of spirit, neither can spirit be independent of matter for its expressions. They so interblend that, the dividing line cannot be detected by the untrained eye of the exact scientist. But, that time is not far distant, when the scientists will prepare and evolve their interior being to take up the spiritual thread, exactly where the visible thread ends, and carry forth the work, as far as the mortal mind of man can penetrate, while embodied in the physical form.

God hasten this day is my prayer, for then man will become more spiritual and aspiring for advancement and knowledge, thus, setting up vibrations that will create higher and loftier conditions for the physical man. Aye! then they will know that, even the birth of the world itself, owes its primal genesis to the desire of the human atom for earthly embodiment.

Here is where exact science, or the counterpart of Alchemy, becomes both profitable and helpful. Says Paracelsus: "The true use of chemistry is not to make gold, but to prepare medicines." He admits four elements—the STAR, the ROOT, the ELEMENT and the SPERM. These elements were composed of the three principles, SIDERIC SALT, SULPHUR, and MERCURY. Mercury, or spirit, sulphur, or oil, and salt, and the passive principles, water and earth. Herein we see the harmony of the two words, Alchemy and Chemistry. One is but the continuation of the other, and they blend so into each other that, they are not complete, apart.

The chemist, in his analysis of the various component parts of any form of matter, knows also the proportional combinations; and thus, by the Law of Correspondence, could, by the same use of the spiritual laws of Alchemy, analyze and combine the same elements from the atmosphere, to produce the corresponding expression of crystallized form. By the same laws, are affinities and antipathies discovered and applied, in every department of Nature's wonderful laboratory.

Chemistry is the physical expression of Alchemy, and any true knowledge of chemistry is:—not the knowing of the names of the extracts and essences, and the plants themselves, and that certain combinations produce certain results, obtained from blind experiments, yet, prompted by the Divine spirit within; but, knowledge born from knowing the why and wherefore of such effects. What is called the oil of olives is not a single, simple substance, but it is more or less combined with other essential elements, and will fuse and coalesce with other oils and essences of similar nature. The true chemist will not confine his researches for knowledge to the mere examination, analysis, and experiments, in organic life; but will inform himself equally, in physical astrology; and learn the nature, attributes, and manifested influences of the planets, that constitute our universe; and, under which, every form of organic matter is subject, and especially, controlled by. Then, by learning the influence of the planets upon the human family; and that special planetary vibration that influences the individual; he can intelligently and unerringly administer medicines to remove disease in man.

A familiarity with the mere chemical relations of the planet to man, makes still more apparent, the mutual affinity of both to the soil, from which they appear to spring, and to which, they ultimately return; so much so that, we have become conscious, that, the food we eat is valuable or otherwise as a life sustainer, in proportion to the amount of life it contains. We are so complex in our organization that, we require a great variety of the different elements to sustain all the active functions and powers within us. Man, being a microcosm, or a miniature universe, must sustain that universe, by taking into the system the various elements, which combine to make up the Infinite Universe of God. Animal flesh is necessary to certain organized forms, both animal and man. When I say necessary, I do not mean an acquired taste and habit of consuming just so much flesh a day; but a constitution, which would not be complete in its requirements, without animal flesh. I am thankful such do not constitute the masses.

Science would say, you only require certain combinations of oxygen, hydrogen, nitrogen, and carbon, to sustain all the activities of the physical body. Apparently, this is true. Upon the surface it is, but in reality it is not; because if it were really true there could be no famines. Science could make bread out of stones, as was suggested at the temptation of Christ in the wilderness. And yet, no one knows better than the academies of Science, themselves, that their learned professors would quickly starve to death, if they were compelled to produce their food from the chemical properties of the rocks. They can make a grain of wheat chemically perfect, but they cannot make the invisible germ by which it will grow, become fruitful, and reproduce itself. They can reproduce from the stones in the street the same chemical equivalents that go to compose gluten, albumen, and starch—the trinity which must always be present to sustain life; but they cannot, by any known process, make such chemical equivalents of these substances, do the same thing. Now, if not, why not? Science cannot answer this. A very mysterious shake of the head and profound silence is the only answer. Ask Science HOW THE PLANT GROWS, what causes the atoms of matter to build up root, stem, leaf, bud and flower, true to the parent species from which the germinal atom came. What is there behind the plant that stamps it with such striking individuality? And why, from the same soil, the deadly aconite and nutritious vegetable can grow, each producing qualities in harmony with its own nature, so widely different in their effects upon the human organism, YET, SO COMPLETELY IDENTICAL AS REGARDS THE SOURCE FROM WHICH THEY APPEAR TO SPRING. There must be a something to account for this, and this something, ancient Alchemy alone can scientifically reveal and expound; and, this knowledge lies just beyond that line which calls a halt to material scientists, and says: "You can go no farther; this is beyond your purview. The end of the material thread has been reached, and unless you can connect it with the thread of the next plane, your researches must stop."

Before entering upon and answering these vital questions, we must digress a little, and make ourselves perfectly familiar with the ideas and revelations of advanced physical science upon the subject, and for this purpose no more trustworthy guide can be consulted than the new edition of "The Chemistry of Common Life," by the late James F. W. Johnson, M. A., England, and revised by Arthur Herbert Church, M. A. In chapter IV on page 56 of this work, upon the anatomy of plant life, we read:

"How interesting it is to reflect on the minuteness of the organs by which the largest plants are fed and sustained. Microscopic apertures in the leaf suck in gaseous food from the air; the surfaces of microscopic hairs suck a liquid food from the soil. We are accustomed to admire, with natural and just astonishment, how huge, rocky reefs, hundreds of miles in length, can be built up by the conjoined labors of myriads of minute zoophytes, laboring together on the surface of a coral rock; but it is not less wonderful that, by the ceaseless working of similar microscopic agencies in leaf and root, the substance of vast forests should be built up and made to grow before our eyes. It is more wonderful, in fact; for whereas, in the one case, the chief result is that, dead matter extracted from the sea is transformed into a dead rock; in the other, the lifeless matter of the earth and air are converted by these minute plant-builders into living forms, lifting their heads aloft to the sky, waving with every wind that blows, and beautifying whole continents with the varying verdure of their ever-changing leaves."

Further on in the same chapter, on pages 62-3, the same eloquent writer continues:

"But the special chemical changes that go on within the plant, could we follow them, would appear not less wonderful than the rapid production of entire microscopic vegetables from the raw food contained in the juice of the grape. It is as yet altogether incomprehensible, even to the most refined physiological chemistry, how, from the same food taken in from the air, and from generally similar food drawn up from the soil, different plants, and different parts of plants, should be able to extract or produce substances so very different from each other in composition and in all of their properties. From the seed-vessels of one (the poppy) we collect a juice which dries up into our commercial opium; from the bark of another (cinchona) we extract the quinine with which we assuage the raging fever; from the leaves of others, like those of hemlock and tobacco, we distil deadly poisons, often of rare value for their medicinal uses. The flowers and leaves of some yield volatile oils, which we delight in for their odors and their aromatic qualities; the seeds of others give fixed oils, which are prized for the table or use in the arts * * * These, and a thousand other similar facts, tell us how wonderfully varied are the changes which the same original forms of matter undergo in the interior of living plants. Indeed, whether we regard the vegetable as a whole, or examine its minutest part, we find equal evidence of the same diversity of changes and of the same production, in comparatively minute quantities, of very different, yet often characteristic forms of matter."

From the whole of the foregoing, we observe the exact position to be the one we have previously stated. If such wondrous things can be revealed to us through the physical science of chemistry, what think you must be hidden from our physical sight and knowledge by the veil which hangs between matter and spirit? Think you not, it is worth the effort to penetrate beyond that point where the atom disappears from the view of the scientist?

If plants produce such wonderful phenomena in their life and influence, what must the Divine organism of man have concealed within his microscopic universe, to study and comprehend? Plant life is merely the alphabet of the complex, intricate, and multitudinous processes, going on in the human body.

And, as the mechanical microscope of physical science cannot reveal the why and the wherefore, let us, for a brief moment, disclose some of the wonders that declare their existence, when subjected to the penetrating alchemical lens, of the inward spirit. The first thing that intrudes itself upon our notice, by virtue of its primary importance, is the grand fact of biogenesis—life emanating from life. We perceive every external form to be the physical symbol of a corresponding degree of spiritual life; that each complete plant represents a complete cycle, state, or degree of interior existence; that it is made up and consists of countless millions of separate atoms of life; that these atoms of spiritual activity are the real instigators of the life and motion of corresponding material atoms; that they ever obey the

Divine impulse of co-operative unity, in their chemical, as well as their spiritual affinity. Consequently, everything in the form of material substance must be, and is, but the means for the phenomenal expression of incarnating spirit; the organism of man, a tree, a plant, or an animal, being no exception to this Divine, omnipresent law of creative life.

To the true Alchemist there can be no mystery surrounding the wonderful phenomena mentioned in the work we have quoted, in plants extracting from the same rocks, soil, and air, qualities so manifestly different— deadly poisons, healing balsams, and pleasant aromas, or the reverse, from the same identical plant foods. Nothing is more wonderful or mysterious, than, the same alchemical processes, which, are hourly being enacted within our own bodies. From the same breath of air and the same crust of bread do we concoct the blood, the bile, the gastric juice, and various other secretions; and distil the finer nervous fluids, that go to build up and sustain the whole of our mental and dynamic machinery. It is the same ancient story of the atoms; each part and each function endowing the same inorganic chemicals with their own spiritual, magnetic, and physical life-qualities, by what appears, to the uninitiated observer, a miraculous transmutation of matter, but which is, in reality, the evolution of organic form from inorganic materials, in obedience to the Divine law of spiritual progression. Who could stop with exact science? For, when we come to consider the apparent mysteries of life and growth by the aid of this alchemical light, the shadows flee, and all the illusions of Nature's phenomenal kaleidoscope vanish before the revelation of the underlying spiritual realities. We know that the plant, being the physical expression upon the material plane of a more interior life, endows its outward atoms with their peculiar qualities. THESE QUALITIES ARE NOT DRAWN DIRECTLY FROM THE SOIL; the soil only becoming the medium for their complete or incomplete expression, as the case may be; i.e., supplying the necessary inorganic atoms. Hence, the deadly qualities of aconite, and the generous life-sustaining qualities of the nutritious vegetable, BEING SPIRITUAL LIFE ENDOWMENTS, conveyed to the material substance, abstracted from the soil and withdrawn from the atmosphere, are no mystery; their effect upon the human organism being exactly that, which is produced by their spiritual affinity or antipathy, as the case may be. And this also shows and explains, why purely inorganic chemical atoms, though they be exactly the same as the organic substances, from a strictly scientific standpoint, YET FAIL TO SUPPORT LIFE, because such chemical equivalents lack the organic spirituality of the interior life, which alone, gives them the power and function to support the same. They fail to fulfill the requirements of the alchemical law of life for the support of life—in other words, biogenesis.

And, too, this inorganic life may be parted from the plant or vegetable, if it be too long severed from the medium which transmits the spiritual life, from the inorganic world to that of organic matter. Vegetables, fresh from the ground, or parent stem, retain this life if at once prepared for food, if not overcooked, which is so often ignorantly done. This is the secret of sustenance from foods. Nature's perfected fruits and vegetables are overflowing with the life-giving essences, and, if eaten direct from the tree or parent stem, that life is not lost, but transmitted to our organisms, and replenishes the wasting system with a living life. Much less of such food is required to completely satisfy and nourish the body than if the life had partly departed or been destroyed.

Briefly stated, then, everything within organic Nature is the expressional symbolic manifestation of spirit; every form being a congregation of innumerable atoms of life, revealing their presence in material states; each organic form, or, rather, organism, evolving under the central control of some dominating Deific atom or soul, which, by virtue of past incarnations and labors in its cycle of evolution, from the mineral up to man, has achieved the royal prerogative to rule within its own state. Man being the highest representative form—the grand finale in the earthly drama—sums up and contains within himself everything below, and THE GERMS OF EVERYTHING BEYOND, THIS STATE. He is truly a microcosm, and represents in miniature the grand Cosmic Man of the Heavens. Every living force beneath him corresponds to some state, part, or function, which he has graduated through and conquered, and which, in him, has now become embodied, as a part of his universal kingdom. Consequently, all things are directly related to him, in the grand universal unity of spiritual life.

This cannot be realized and comprehended by the physical man, nor conveyed to his outer senses by the physical sciences. He must bring into active use the inner man, the real being, which inhabits and controls the outer organism, and through its instrumentality, understand the interior source and workings behind the phenomena of manifested being. So we see that, exact science cannot take us far, yet, it is a mighty factor, in the evolution of the microcosm Man, and in consciously relating him to the Infinite Macrocosm—God, Spirit, All.

CHAPTER VIII. ALCHEMY-PART II

Paracelsus, the most celebrated of the alchemists of the Middle
Ages, thus mystically speaks of his art:

"If I have manna in my constitution, I can attract manna from heaven. Melissa is not only in the garden, but also in the air and in heaven. Saturn is not only in the sky, but also deep in the ocean and Earth. What is Venus but the artemisia that grows in your garden, and what is iron but the planet Mars? That is to say, Venus and Artemisia are both products of the same essence, while Mars and iron are manifestations of the same cause. What is the human body but a constellation of the same powers that formed the stars in the sky? He who knows Mars knows the qualities of iron, and he who knows what iron is knows the attributes of Mars. What would become of your heart if there were no Sun in the Universe? What would be the use of your 'Vasa Spermatica'[*] if there were no Venus? To grasp the invisible elements, to attract them by their material correspondences, to control, purify, and transmute, them by the ever-moving powers of the living spirit—this is true Alchemy."

[*] Astral germs of subjective life forms:—it is the latent, "to be".

Thus, in a very few simple words, we find this master of the art revealing the whole arcana of that mysterious science, which has for its chief object and goal, the discovery of the "philosopher's stone," which confers upon its fortunate possessor the blessings of immortal youth. Therefore, we cannot possibly do better in the commencement of our present study than, to minutely examine each particular sentence and endeavor to discover his true meaning, which, like all mystical writing, is so apparent, yet cunningly concealed, as to excite the student's admiration.

"If I have manna in my constitution, I can attract manna from Heaven." The manna here spoken of does not specify any particular thing, but is of universal application, and is simply used as an unknown quantity, like x, y, z in mathematics. But, ever since the days of Paracelsus, half-initiated mystics and bookworm occultists, have endeavored to discover what this manna really was. Some, the more spiritual, were of the opinion that, it was spiritual power, or purity of spirit; others imagined it to mean special magnetic qualifications, similar in nature to the so-called gifts of modern spiritualistic media. The concealment of the truth is unique, and consists in its very simplicity; and, when correctly expounded, should read: "I am the microcosm, and all the visible and invisible universe dwells within me, so that whatsoever power I have in my constitution, I can attract its correspondence from Heaven." Paracelsus must have smiled to himself when he wrote "If I have manna," etc., because his whole writings strive to prove man the miniature of Deity. Further along, he explains himself by pointing out the real Law of Correspondence, thus: "Melissa is not only in the garden, but also in the air, and in Heaven. Saturn is not only in the sky, but also deep in the ocean, and Earth." The illustrations are beautiful, The life of the plant, the "anima floralis," pervades the atmosphere and the interior states of spiritual life, where it becomes in the highest degree beautiful, and beneficial to the soul. A reference upon this point to "The Light of Egypt" Vol., I, may not be considered out of place. Upon page 74 it is written: "The flower that blooms in beauty, breathing forth to the air its fragrance, which is at once grateful to the senses and stimulating to the nerves, is a perfect specimen of Nature's faultless mediumship. The flower is a medium for the transmission to the human body of those finer essences, and of THEIR SPIRITUAL PORTION TO THE SOUL; for the aroma of the flower is spiritualized to such a degree as to act upon the life currents of the system, imparting to the spiritual body a nutriment of the finest quality."

Thus, here is where the knowledge of the alchemical attributes of plants, as applicable to man, can be most beneficially utilized. Plants and flowers, whose attributes and aromas harmonize with the complex organism of man, should be selected for the house and garden, for, they are mediums to transmit the finer essences and aromas to the spiritual constitution of man; the plant to the physical, and the aromas and essences of the flowers to the soul.

Antipathies in plants and flowers would bring a similar evil influence, as the discords of the antagonistic human magnetism. It would not be so apparent, but more subtle, yet nevertheless effective in result.

Our attention is next drawn to the planet Saturn, which, we are informed, is not only shining in his starry sphere of the heavens, but is also buried in the ocean depths and embodied in the stratas of the earth. It is almost needless to add that, our author refers to those substances naturally Saturnine in their quality of life and expression, such as lead, clay, and coal, among the minerals, and various deadly plants among the flora, the chief of which is the aconite or monkshood, so significant of Saturn and the isolated, monkish hermit. After some repetition, in order to impress the truth of correspondences, our author exclaims: "What is the human body but a constellation of the same powers that formed the stars in the sky?" Truly, what else? for, "he who knows Mars knows the qualities of iron, and he who knows what iron is knows the attributes of Mars." Could anything be plainer? We think not.

From the foregoing, which a long experience and much critical investigation and research have demonstrated as true, we cannot avoid the conclusion that Alchemy, equally as well as every other science, religion, or system of philosophy formulated by man, resolves itself, ultimately, in all its final conclusions, into the one universal parent of all wisdom.

ASTROLOGY, the Science of the Stars, in unison with the Science of the Soul, was, and still is, the one sublime center of real learning. It constituted the sacred fountain of living waters, from whose placid depths there rayed forth the Divine revelations of man, his whence, where, and whither; and under the careful conservation of a long line of gifted seers, it shone forth to the sons of men, as the sacred Hermetic light in the Astro-Masonic wisdom of Egypt's ancient priesthood.

It is not lost to us to-day. The same book lies open before us that faced our ancient forefathers. It is standing out clear and distinct, waiting to be read by the sons of men. We can learn its language, and from its pages, we ourselves can read our relation to God and our fellowman. Shall we not heed the whispering intuitions of the soul and place ourselves in conscious rapport with the whole?

This sublime Book of Wisdom was written by God Himself, to convey to His children the knowledge of His powers, attributes, and relation to all creative life. We cannot see that Divine Spirit which we call God. No; but as long as the finite form exists as such, we will have the spirit's manifestations to learn from. Never will the Book of God be closed to the searching eye of the soul. There will always be presented to his vision lessons to study, and practical experiments to perform, to lead the soul into deeper mysteries. Until man fathoms his own universe, he cannot understand God. "Know thyself" is as applicable to-day as when the famous, immortal and mystic utterance was inscribed on the porch of the temple at Delphi.

Before this wonderful, divinely elaborated, but complex system can be fully realized, it is necessary that the student should comprehend, very distinctly, the two states of existence, the internal and the external, and become familiar with the laws of correspondences. And it seems strange that of all Sciences, that of medicine should have so completely failed to grasp this living truth, since every atom of medicine administered, invariably acts upon this alchemical principle. When the human organism has become discordant in some of its parts, it is because the interstellar vibrations have aroused various states within the human kingdom into a condition of rebellion against the supreme will. Man's ignorance favors such seditious movements, and his general habits and code of morals stimulate them to undue activity. The final result is disease—disorganization of the parts and functions, and those medicines corresponding TO THE SAME FUNCTIONAL DEGREE OF LIFE WITHIN THE GRAND MAN, cure the disorder, when administered properly and IN TIME, whereas, if given to the perfectly healthy organism, THE ATOMS PRODUCE SIMILAR

SYMPTOMS TO THE DISEASES THEY ALLEVIATE, because it is their mission to either subdue or be subdued, and when disease prevails the medicinal atoms, acting in unison with the natural parts and functions they affect, conquer or subdue the inharmony, and vice versa, as before stated. In all cases of disease and medicine, it is a simple question of A WAR BETWEEN THE ATOMS, and, therefore, the most potential forces within Nature are always at the command of the true Alchemist, because he knows bow and when to select his fighting forces, and when to set them in motion, for the best results.

Hahnemann, the founder of the Homeopathic system, has approached THE NEAREST to this alchemical truth, and as a consequence, we find it is in actual practice, the most natural, scientific, and successful system of medicine, yet given to the world; based, as it is, upon the well-known law of affinites, "Similia similibus curantur," "like cures like," being a very ancient axiom in the astrological practice of physic.

Bulwer Lytton, who had become thoroughly convinced of the great value and importance of uniting ancient Alchemy with modern medicine, makes the hero of his immortal story declare: "All that we propose to do is this: To find out the secrets of the human frame, to know why the parts ossify and the blood stagnates, and to apply continual preventives to the effects of time. THIS IS NOT MAGIC; IT IS THE ART OF MEDICINE, RIGHTLY UNDERSTOOD."

It is a fact that, the molecules of the body are all changed within twelve months; that every cell in the human organism is born and grows to maturity within that space of time. Nature is absolutely impartial. She draws from the atmosphere that she may reproduce a fac-simile of everything she finds upon the surface of the body. So, if there be a sore, or festering ulcer, the atoms which are thrown off attract similar atoms, so as to reproduce the ulcer or sore, and thus prevent the disease from getting well of itself until it has worn itself out.

Further, every vein and canal throughout the entire body, from youth to maturity, is being coated with carbonate of lime, or lime in some form. The coating of the walls of the veins in such a manner, prevents the free circulation of the living matter; then, the real vitality of the food which we eat, is simply passed off through the pores, or through the bowels, or through the system, because it is unable to penetrate through the lime.

If that prevention which produces old age can be attained, then physical youth will continue.

The first step to take is to dissolve the lime in the body. Drink nothing but distilled water, in either tea, coffee, or any other form, and drink freely of the sweet juices of the grape and apple.

The food that we eat contains lime in a living form, and it is the living lime we need to build up the living bones, for the lime and the magnesia that we take in the water is crystallized dead mineral, possessing no responsibility of life, and the lime in our food is quite sufficient for all purposes. For everything we take in excess, Nature makes us pay the penalty.

The first principle of long living is to keep all channels of the body perfect and free from coatings of lime.

The second is that of youthful ideals of the mind. The soul never grows old.

The third principle is dynamic breathing, which is storing up the oxygen in sufficient quantities, to supply the tissues with sufficient fuel, for combustion.

These three principles, acting in unison, contain the true basis of physical life and a means of long living. Old age is simply the petrifaction of the body through lime, and the incorporating of erroneous thoughts into the organism.

It is the true Alchemy of human existence, and the preventives, in each and every case must contain the spiritual correspondence to the cause they seek to remedy; and, though the followers of Hahnemann base the whole of their procedure of treatment upon their master's fundamental law of "Similia similibus curantur," yet, there may be a few rare cases wherein, this undeviating method would not apply with the required effect. In such a case, the Alchemist would resort to the well-known law of opposites, and base his treatment upon the dogma of "Contraria contrariis curantur," so long the pet theory of the Allopathic school. They work upon the hypothesis that, like attracts like, and, if disease exist, those elements must be administered to set up the vibrations that will produce the polar opposite. If the body was racked with pain, those medicines would not be given that would create or increase similar conditions, but, their antipathy would be introduced into the system or applied locally to extinguish the foe.

So long as mankind remain within the semicrystallized state of soul development, so as to require the aid of external forces to support the human throne within its earthly temple, mercenary troops will exist to supply these supposed supports.

Unquestionably, the astrological law is the true system of medicine, which treats disease by sympathy or by antipathy, according to the nature of the case, and the efficacy of the remedies at hand. This method is the only natural one, and has been thoroughly demonstrated by the numerous "provings of drugs" under Hahnemann's law.

Happily, the time is not far distant when, the incarnated spirit will be able to use its own slumbering forces, and subdue all suffering and symptoms of disease in their very first inception, by virtue of its purer life and the dynamic potencies of its own interior, spiritual thought. Already, mental therapeutics is taking an advanced position among liberal, progressive minds, and nothing demonstrates so clearly and forcibly the grand, alchemical law of life-growth and decay, as the imponderable, invisible forces, which, constitute the materia medica, or remedial agents, of mental, magnetic, and spiritual healing.

Perhaps the most recondite subject connected with the healing art divine is, the modus operandi of medicinal action, upon the human body. A subject so simple and self-evident to the Alchemist, remains a profound mystery to the educated physician of the medical college; so much so that, we are tempted to ask of them: "Can you explain the modus operandi of drugs?" Dr. William Sharp, one of the most advanced physicians of the Homeopathic school, in one of his well- known "Essays on Medicine," says: "In respect to the manner of action of drugs we are in total darkness, and we are so blind that the darkness is not felt. KNOWLEDGE OF THIS KIND CANNOT BE ATTAINED; it is labor lost and TIME wasted to go in search of it. True, hypotheses may be easily conceived; so may straws be gathered from the surface of the stream. But what are either of them worth? There is this difference between them—straws may amuse children, and hypotheses are sure to mislead physicians."

It is when the Occult Initiate observes to what helpless conditions the practice of medicine has fallen, that, he would, if be could with any possibility of success, implore the angelic guardian of the human race to open the spiritual sight of men, that they might see, as he sees, the Divine relationship, and spiritual correspondence, of everything in the wide universe to man.

Nature's laws move slowly and imperceptibly, yet surely and exact, and the time will certainly come when man will be forced into consciousness of these laws, whether he will or no. Nature is no respecter of persons, and those who will not move and progress, in harmony with her laws of advancement, must, of necessity, pass out with the old.

Alchemy, as it relates to the healing art, is the most noble in its object and beneficial in its effects, of all the many subdivisions of the sciences, because, it alleviates the pains and morbid afflictions of suffering humanity. We have given quite sufficient of its astrological aspect in the second part of "The Light of Egypt," Vol. I, wherein the four ancient elements are translated into their chemical correspondences of oxygen, hydrogen, nitrogen and carbon, which still constitute the four primary elements of the most advanced chemistry to-day. They enter more or less into every organic form and substance, which is known, in various combinations and proportions. The human organism is principally composed of them; so, likewise, is the food that supports physical life, and the air we breathe is but modifications of the same atoms.

As man's constitution embraces a microscopic atom of all the essences and elements, corresponding to the whole; so does the air; and much, that we depend upon our food to supply, can be extracted from the atmosphere by breathing. Every breath we breathe is new life, or death.

Herein is the secret of success or failure, in certain localities, and under certain conditions. If we have iron within us, could we extract or attract iron from Saturn's district? Or, if the element within us could attract gold, could we obtain it from the coal fields?

Therefore, it is only natural that the medical remedies we employ to restore the organism, when afflicted with disease, should group themselves into similar correspondences, and so, in a general sense, we find them; for we note that the brain, the circulation, the lungs, and the stomach, are the four chief citadels of the body; the heart, of course, representing the center of circulation. And this also explains, further, if that were necessary, why the principal remedies of the homeopathic system are so speedy and direct in their action. The four principal drugs, which stand as representatives of their class, are aconite, belladonna, phosphorus, and pulsatilla. These represent the quadrant, for light is not more nicely adjusted to the eye, nor sound to the ear, than aconite to the circulation, belladonna to the brain, phosphorus to the lungs, and pulsatilla to the stomach; while ramifying in the seven directions indicated by the seven primary planets, we find stimulants, tonics, narcotics, nervines, alteratives, cathartics and diuretics, as the natural material correspondences thereof.

That we assign phosphorus to the lungs may appear startling to the orthodox student, especially when, he calls to mind the fact that phosphorus has long been recognized in medical science as a brain food and medicine. Anticipating such mental questions, we reply that in medicine, from the alchemical view, we are occupying a wholly different standpoint; i.e., the power of controlling the functional action of the body, in this view of the case, and the fact that, the lungs and the brain are in the most perfect affinity, there will remain no mystery upon the subject.

The Alchemy of stones and gems attracts our next attention. Affinites and antipathies to the human constitution, are to be found in these crystallized representatives of the subtle, invisible influences emanating from our planetary system. They are the mediums for the transmission of corresponding attributes and influences of existing powers and potencies, and if carried or worn upon the person, they will bring the person in direct rapport with the invisible forces within the universal system.

Here again Hahnemann's scientific philosophy would prove effectual, that "Similia similibus curantur." Would the fiery influence of a topaz attract much from the realms of a chrysolite? Or, the crystallized, airy forces of a sapphire be a suitable medium for the earthly forces of a jasper?

Gems and stones are dead or living realities. They live, slumber and die, and have their potent existence as do the organic forms of matter. They are, usually, imbued with the vivifying spark of Divinity, and shine forth and exert their influence through the magical powers attracted to them from the forces of Nature. A real, living entity abides within them that can be seen by the clairvoyant vision, and to the trained student in Occult lore, this entity can be made to become an obedient servant, giving warning of the approach of danger, impressions of men and things, and warding off discordant influences surrounding us; or that, which we may contact from the magnetic and personal environments in our relations in the social world; or that which may be projected to us from the invisible realms of life.

Think you the pryamids would be intact to-day, if the stones from which they were built had been promiscuously selected? They were chosen by Adepts in the knowledge of the Laws of Correspondence and antipathy and affinity. The sphinx also stand as monuments to the heights of wisdom that man can attain.

Metals also can be followed out on the same lines as the gems and stones.

Much as we would like to continue, we are compelled to bring this discourse to a close, even though in doing so we must of necessity omit much of vital interest to the student. We will, therefore, only add that the seven basic metals stand as the crystallized representatives of their respective groups: Gold for the Sun, Silver for the Moon, Tin for Jupiter, Copper for Venus, Quicksilver for Mercury and Lead for Saturn. Each finds it own sphere of action within the temporary abiding place of the human soul on earth—the physical body. So, likewise, the twelve constellations and their corresponding talismanic gems, representing in their glittering array the anatomical Zodiac of the human frame, and typifying the spiritual quality of the atoms, there congregated, in every degree of life. These, and a thousand other mysteries, had we the time, might be unfolded to the student's view with considerable advantage, but we are compelled to refrain. The philosopher's stone is near at hand. Seek it not in remote spheres or distant parts of the earth, for it is ever around you and within, and becomes the golden key of true wisdom, which prepares the soul for its higher life and brighter destiny. It is the still, small voice of the awakened soul, that purges the conscience from suffering, and the spiritual body from earthy dross. It is that, which treasures not the corrupting, delusive wealth of Earth, nor the transient powers of mammon, but garners the fruits which spring from the pure life, and treasures the jewels of heaven. Vainly will you seek for this stone of the wise philosopher amid the turmoils, sufferings, and selfishness of life, unless you accept your mission upon earth as a duty, delegated to the soul, from Heaven. Eschew the evil thereof, and hold fast that which is good. To do this, means to expand with the inward truth and become one of the "pure in heart," in which blessed state, the magical white stone, conveying A NEW NAME, reveals the living angel within, to the outward man. Then, and then alone, doth he know the Adonai.

Such are the Divine, spiritual principles upon which the higher Alchemy of life is based. They seek only to establish a Divine, conscious at-one-ment between the angel, the man, and the universe, and to this end, we conclude with the words of the immortal Paracelsus:

"To grasp these invisible elements, to attract them by their material correspondences, to control, purify, and transmute them by the ever-moving powers of the living spirit, THIS IS TRUE ALCHEMY."

CHAPTER IX. TALISMANS

Words are the symbols of ideas, and bear the same correspondence to the physical brain as matter does to spirit, a medium of expression, and are subject to continual change in their application and meaning, in exact proportion to the changing mental and moral condition of the people. As the planet, as well as man, is continually progressing, so must there be a higher and nobler conception of ideas. Hence, words or expressions must change, to convey the progressive spirit, that is constantly taking place. Therefore, it is always interesting, as well as valuable, for the Occult student to go to the root of each word connected with his philosophy, in order to learn the real sense in which the word was used by the ancients, from whom his mystic lore has descended. The true meaning, as well as the words themselves, have become as mystical as the lore itself. Hence, each student must commence as a beginner in any foreign language, which he does not at present understand. In following this method of procedure he will, at least, escape the dense and interminable confusion of modern opinions upon subjects of which the writers thereof, are partially or wholly ignorant.

No better illustration of this can be afforded than by the word "Talisman," derived from the Greek verb "teleo," which means, primarily, to accomplish, or bring into effect. But, in its real, and therefore higher, sense, it means to dedicate, consecrate, and initiate into the arcana of the temple mysteries. But, in the present day it means a piece of imposture, connected with some magical hocus pocus of the ignorant and superstitious mind, a vulgar charm, that is supposed to bring the owner thereof some material benefit, irrespective of his mental, magnetic, and moral condition, "and," says the learned Webster, after describing his idea of such things, "they consist of three sorts, astronomical, magical and mixed." But in what sense the "astronomical" differed from the "magical" we are not informed, nor is any light thrown upon the peculiar nature of that class designated as "mixed." In fact, the lexicographer so mixes up his definitions that, we are unable to distinguish anything in particular, but his own individual ignorance.

So it has become, in every branch of learning. Words and their meanings have become so mixed in their use and application, that, the world is full of discords and misunderstandings, which lead into dissensions and contention, among all schools of thought, sects, and isms; and lastly, though not the least serious, it has reached into the close relations of the human family.

All writers and speakers, as well as the readers and listeners, should acquaint themselves with the derivation and meaning of words.

The fact stands very clearly defined that, Talismans are confused in the minds of the present generation with magical charms, which depend for their effects, upon the power of the idea or thought, which the formulating magician impresses upon the substance of which they are composed. If the magical artist be expert, and endowed with an exceedingly potent will, his charm may become very powerful, when worn by the person for whom it was prepared. But, if this one grand essential be lacking, no amount of cabalistical figures and sacred names will have any effect, because, there can be no potency in symbols apart from the ideas and mental force they are capable of arousing in the mind of the maker. Solomon's Seal is no more powerful, when drawn upon virgin parchment, with a weak will, or in a mechanical state of mind, than a child's innocent scribbling upon its slate. But, if the artist realizes the mysteries symbolized by the interlacing triangles, and can place his soul en rapport with the invisible elements they outwardly represent; then, powerful effects are often produced.

I am sorry to say that, the knowledge of charms is not confined to the creation of beneficial talismans. Its perversion has led to the diabolical practices of the Voodo and Black Magician, whose work is wholly, either for gain or revenge. Nothing, but the most extreme selfishness lies beneath such immoral practices, but, as there must be a light to reflect a shadow, so a charm must follow a talisman. Magical charms, then, are simply natural objects, possessing but little active virtue in themselves, but, owing to the mediumistic nature of their substances, are endowed with artificial powers, of temporary duration, by virtue of the idea and thought impressed upon them, through the mental magic of the maker; and in this sense, a charm must be clearly distinguished from "teleo," the Talisman. The very names suggest their difference, and, above all other men, students in Occultism should strive to become thoroughly educated in the true sense of the term, MEN OF LETTERS, by virtue of (as Ruskin calls it) "the kingship of words." "Charm" is derived from the Latin "carmen," a song that fascinates, and means to control by incantation, to subdue; while Teleo concerns the secret powers and wisdom of consecration and initiation. It is because of modern misuse of antique terms that, we have considered this somewhat lengthy explanation necessary, in order to clear away the accumulated debris of the ages, from the true foundation of our present study.

A Talisman is a natural object, containing the elemental forces of its own degree of life, in a state of intense activity, and capable of responding to the corresponding quality of life, OUTSIDE OF ITSELF, that emanates from the same spiritual state, either by sympathetic vibration or antagonistic currents, the nature, power, quality, and degree of life, which the various natural objects represent, being a part of the temple curriculum of initiation. Hence, the name, by which the latent power of these natural objects became known, was in strict harmony with the facts involved.

In order to prevent any possible misconception upon the subject, let us briefly restate the definition in a different way: A Talisman is the exact antipodes of a charm. This latter is the artful and temporary result of man's mental power; the former, the natural production of universal Nature, and as permanent and enduring as the substance of which it is composed, DURING THE PRESENT CYCLE. And yet in some sense, it may be quite correct to say that, a Talisman ACTS LIKE A CHARM, and vice versa, that charms ACT LIKE A TALISMAN, providing that, the real vital difference between them, is maintained in the statement.

Now that we have our subject clearly defined, let us carefully examine HOW AND IN WHAT SENSE a given natural object becomes Talismanic, for it must appear self-evident to all that, one and the same substance cannot constitute a Talisman for everyone, and for everything. They must naturally differ, as widely in their nature and quality, as mankind differ in physical, mental, moral, ethical, and temperamental, development. And, yet, though, man may so differ from his fellow man; the ignorant Esquimau, killing seals in his kayak, may belong to the same spiritual quality of life as the Harvard professor, who obtains his subsistence by daily discourse upon the sublime harmony of the infinitely small with the infinitely great, throughout the manifested universe of matter, and wherever we find this KINSHIP of the spirit, we shall find the same identical Talisman acting alike upon each, whenever they shall come en rapport with it. Mental, moral, and physical development, never alter the real nature of the internal man. Culture only brings to the surface, into active use, the latent possibilities lying concealed within the human soul. It only allows him to exercise his functions upon different planes, and with different effect.

Every natural department of Nature corresponds to some peculiar specific quality and degree of life. These have been divided, for the sake of convenience, into four primary groups; and each group again subdivided into three, corresponding to the four cardinal, four succedent, and four cadent houses, of the astrological chart; therefore, the twelve signs of the Zodiac; these constituting the Cycle of Necessity within physical conditions, wherein, the ever-measuring or decreeing tidal flow of life from solar radiation throughout the year, represents the twelve groups of humanity, of lower animated Nature, of vegetation, and crystallized gems. Every human being is ushered into the world under the direct influx of one or more of these celestial divisions, and by virtue of the sign occupying the horizon at the moment of birth, absorbs such influx, and becomes endowed with a specific polarity, by virtue of which, lie ever afterward, during such expression within physical conditions, inspires with every breath, that specific life quality from the atmosphere, corresponding to the same degree of the universal spirit. Consequently, that gem, or those gems, representing and corresponding to HIS HOUSE OF LIFE, become to him, a Talisman, because of their relationship—their spiritual affinity. These are all given in the second part of Vol. I. THE METALS never become Talismanic, because of their comparatively negative degree of life, and for this reason also, they make the most powerful charms. Certain combinations of metals, and in proper proportions, increase the potency and magnetic influence of a charm; and here, too, the laws of antipathy and affinity come into practical use.

A true expert will know his metals, or metal, and his client, before commencing his magical work.

Those persons who derive most virtue from a Talisman are those who belong to the most sensitive, or interior state, within such degree of life, and who are dominated by one sign only. Thus, if we find one sign occupying the whole of the House of Life, or practically so, as when the first face of a sign ascends, we may be sure, other things not interfering, that such a native will receive great benefit from wearing its Talismanic gem. If a person of good intellectual powers and sensitive spirituality, be born when the lord of the ascendant occupies the RISING SIGN, as, for instance, Mars in Aries, or Sun in Leo, we may be sure that, the Talismanic gem, in their case, will be exceedingly powerful, because, all the Astro-physical conditions are then most favorable for the expression of natural forces, and, if worn upon, or near that part of the body which the sign rules, the power and influence is more powerful and beneficial.

In wearing them, take them to you as a part of yourself, a part of your higher self, a thing to be heeded, listened to and obeyed. They will usually make their presence most pronounced when something arises to disturb the harmonious vibrations that naturally and quietly go on between the person and the interstellar spaces above. They are like the sensor and motor nerves—they never make their presence known, except, when danger encroaches.

Having explained in what sense gems become talismanic, we have now to disclose the modus operandi—THE HOW.

The gems contain the life quality of their own astral nature. Man, as a higher expression, only, of the same universal biune life, contains the same. Like two electric currents, MAN, THE POSITIVE POLE (comparatively), attracts unto himself THE MINERAL LIFE OF THE GEM, which thus, becomes the negative pole. A complete circuit is formed and maintained, as long as they remain in contact. Gems belonging to a different quality of life, not being en rapport with his astral state, have no good effect, because, no current flows between them. Thus, the Talisman acts in unison with the psychic, or soul-principle, of man, aiding the organism to sustain health, stimulating the mental perceptions, and spiritual intuition, and affording in a remarkable manner, many premonitions of coming danger, when the individual is sufficiently sensitive to perceive them. And now, per contra, as there are gems that act in sympathy with man, there must be, and in fact are, gems that act upon contrary principles; i.e., antagonistic, and these belong to purely antagonistic elements, as Air to Earth and Fire to Water, unless the native be born under BOTH forces, as Mars in Cancer rising, or the latter part of one sign and nearly the whole of another of an opposite nature, occupying the ascendant. Such natives are pure neutrals, and such might wear the gems that belong to the most powerful planet of the horoscope, or that triplicity holding the most planets; then, they are usually combined, the planet and the triplicity.

There are, of course, innumerable substances, more or less, capable of talismanic virtue to particular individuals. But those gems, and similar ones, that are given in "The Light of Egypt," Vol. I, are the most powerful. To these may be added the opal, under Scorpio; the garnet, under Aries; and the turquoise, under Cancer, when Saturn is therein; and the aquamarine, under Pisces; and among the temporary talismans of vegetation we may add that, the young shoots, bearing the flower and seed vessels, are the portions of chief virtue, and the young shoots of trees. These are often used in locating mines, wells, oils, etc., that lie hidden beneath the surface of the earth, and in the hands of a negative, sensitive person, seldom fail to reward the searcher with success. These should always be gathered when their ruling correspondences are rising, or, BETTER STILL, CULMINATING UPON THE MERIDIAN. These will be explained in the chapter on The Magic Wand.

We have now reached the limits of our present study, and have only to state that all gems, like the human organism, are in one of three conditions: alive and conscious, asleep and UNCONSCIOUS, or dead and powerless. These conditions can only be discovered, in stones, by the trained lucid or the instructed neophyte. Stones that are sleeping require to be awakened. This, also, can only be done by the trained student or Adept. Those that are dead, are USELESS as Talismans, no matter how beautiful they appear as ornaments.

Gems and stones are also sexed, and those who wear them would receive the best effect if they should wear those of opposite sex, although either is powerfully potent in their influence upon the individual. How very ignorant the children of men are, of the subtle, silent, yet obedient servants, that everywhere, surround them. Here, again, that Divine spark, which lies embedded within the crystallized forces of Nature, is exerting its subtle, spiritual influence, in making man's very selfishness, and love of ornament and show, a means, to bring forth these silent monitors, knowing ere long that, their true power and potency will be known, and consciously utilized by him, as potent factors in his soul's evolvement and physical development.

The twelve representative gems within the cold stratas of matter, stand as the material representatives of their stellar counterparts in the sky, and constitute the beautiful, glittering, but crystallized, Zodiac of man's physical anatomy.

CHAPTER X. CEREMONIAL MAGIC

The above title has been selected, chiefly, because, in most works treating upon magic we find it wrongly used, and therefore, take the opportunity of explaining the matter, for, there were no such terms in the vocabulary of the ancient Magi.

It is unfortunate, that, words of ancient origin are not more carefully used, and that, we should attach so many different meanings to the same word. The terms "ceremony" and "ceremonial" are nothing more nor less than, what that eminent critic, John Ruskin, would designate as "bastards of ignoble origin," which, somehow or another, have usurped the places of "rite" and "ritual." The word "rite" has descended to us from the Latin "ritus" of our Roman ancestors, and they received it from the more ancient "riti" of the Sanskrit, the Greek equivalent of which is "reo," and means the method or order of service to the gods, whereas, "ceremony" may mean anything and everything, from the terms of a brutal prize fight to the conduct of divine service within the church. But, no such chameleon-like definition or construction can properly be placed upon the word "rite," for it means distinctly, if it means anything at all, the serious usage and sacred method of conducting service in honor of the gods, or of superiors, and requires the attendance of the prophet or priest, or some one duly qualified to fulfill such sacred functions for the time being. The ritual of magic, then, is the correct title of this present study, and as such, we shall, henceforth, term it as we proceed with the course.

Man is especially, and above all creatures, an organizing force, and when to this fact, we add the most interior and powerful of his sentimental instincts—veneration for the powers that be, and for the higher, invisible forces of Nature, his "religiosity," as it has been aptly termed, we cannot wonder that, the earliest races of which we possess any record are chiefly distinguished for their imposing and elaborate religious rites. In fact, it is to the stupendous temples and a colossal sacerdotalism, that, we are indebted for nine-tenths of the relics and records which we possess of them. So true is this that, from what we have been able to discover, we are quite justified in asserting that the ancient races were, above all other things, a profoundly religious people. The temple was the center around which revolved all their genius and art, and the sacred edifice became their grandest achievement in architecture, and its high priest the most powerful individual in the state. In fact, it was in consequence of the real power invested in such sacred office that it was so intimately connected with the throne, and why royalty so frequently belonged to the priesthood or exercised priestly functions. And there can be no real doubt, but that, amongst the pastoral and more spiritual races of Earth's earliest inhabitants, the priest, by reason of his superior wisdom, was the first law-giver; and, by virtue of his sanctity of person and elevation of mind became their first, primitive king, a patriarchal monarch, whose scepter and symbol of power was the shepherd's peaceful crook; just as among the ruder nomads of the inhospitable North, we find the greatest hunters invested with the dignity of chief, whose significant symbol and scepter of royalty, upon their Nimrod thrones, was the trusty, successful spear. And the times in which we live have bad their full effect upon these symbols, so significant of rule. The monarch has transformed the spear into the less harmful mace, while the Church has added an inch of iron to the crook. Therefore, the former has become less war-like, and the latter less peaceful, and, verily, in actual life we find them so,

The patriarchal sire, head of the tribal household, was the original priest; and the hearthstone the first altar around which the family rites were performed; and from this pure and primitive original have been evolved, through progressive ages, the stately temple and the sacred person of the despotic pontiff; from the sincere prayer the pure aspirations of the human heart and the joyous offerings of fruits and flowers to the invisible powers around them; and from the souls of their beloved ancestors has arisen the costly and complicated ritual of theology. And, if the theologians of to-day really knew the lost, secret meaning of their complicated rituals, and the unseen powers lying behind their external symbols, their anxieties for the continued life of their dying creeds would be turned to new hopes and faith, which could be demonstrated to their equally blind followers; that, that which they were teaching they knew, and could practically use the knowledge given forth in their sanctuaries; and, instead of offering up their supplications to an imaginary, personal Deity, their words, rites, and ceremonies, would take on the form and power that such should command, and they would become truly, what their title really means, a doctor of the soul. Then could they, intelligently, lead and direct the souls of their followers to the path of Christ (Truth), which leads up to salvation; not a vicarious atonement, but gaining the at-one-ment through the individual soul's development to a conscious relation, to that Divine spirit, we call God, where it can say "I know."

Out of those simple gifts, which were the spontaneous offerings of loving remembrance and unselfish charity, have grown the prayers, penances, sacrifices, and servile worship, of sacerdotalism. Out of the paternal consideration and love of the aged sire has evolved the haughty, chilling pride of the selfish, isolated priest, and which reflects its baneful influence upon the worshipers at their feet. They have also changed their once sacred, faithful, and reverent, obedience into suspicion and distrust, and with the educated to utter disgust. The light has been extinguished, and priest and people alike are groping about in darkness.

It is strange, yea, passing strange, the amount of human ignorance and folly that is revealed. When we look upon this picture and then upon that, verily we cannot help but ask the question, is mankind really progressing? We know that it is; we are keenly alive to the truth that the Anthem of Creation sounds out "Excelsior"—"move on," but how, and in what way (SPIRITUALLY) we fail to comprehend. The cyclic development of the human soul is an inscrutable mystery.

All the considerations above presented must be thoroughly weighed and understood in order to arrive at the true value of "the dogma and ritual of high magic," as Eliphas Levi terms it; because, amid the vast array of tinselled drapery, the outcome of man's vain conceit and bombastic pride, we shall find very little that can be considered as vital and really essential to the rites of magic. The show, the drapery, the priestly ornaments and instruments, are to the really spiritual Occultist, but, as sounding brass and a tinkling cymbal. That they had, and still have, their legitimate uses, is true, but these uses do not concern magic, per se, nor its manifold powers. They awed the popular mind, and impressed upon the masses a due reverence for the powers that be. They were instrumental in holding the untrained passions of the common herd in check, by a wholesome fear of summary vengeance from the gods, so that this pageantry of magic, the outward priestly show, was more of a politic development than a spiritual necessity, an astute but, philosophical method of enabling the educated few to govern the uneducated many. And it was only when the educational and initiatory rites of the temple became corrupt, and the priest became the persecuting ally of the king—when, in real fact, the priest lost his spirituality in the desire for temporal power and place, that the people began to disbelieve his professions and rebel against his tyrannical control.

The powers that be, are now wielding their sword of justice, and unfurling the knowledge of freedom and truth to the aspiring mind of man. He has begun to feel his bondage and the yoke of oppression. The words of promise and love, instead of lifting him up to the God he has been taught to worship, bow him down in slavish obedience to his priest. Mankind cannot remain in this mental and spiritual darkness much longer. Already I see the break of day, the dawn of a new life, a new religion;

or, rather, the re-establishing of the true, which is as old as Time itself. There is but One Law, One Principle, One Word, One Truth and One God.

The original requirements for the office of priest, and the rites of magic, were, as shown, a primitive, i.e., pure mind; one that had outgrown the lusts and passions of youth, a person of responsibility and experience; and even to this day the priest of the Roman Church is called by the familiar title of "father." And as Nature does not alter her laws and requirements in obedience to the moral development of the race, we may rest assured that the same requirements, of ten thousand years ago, still hold good to-day. You may enter your magic circle, drawn with prescribed rites, and you may intone your consecrations and chant your incantations; you may burn your incense in the brazen censer and pose in your flowing, priestly robes; you may bear the sacred pentacles of the spirit upon your breast and wave the magic sword to the four quarters of the heavens; yea, you may even do more—you may burn the secret sigil of the objurant spirit; and yell your conjurations and exorcisms till you are black in the face; but all in vain, my friend—all in vain. It will prove nothing but vanity and vexation of spirit unless the inward self, the soul, interblends with the outward Word, and contacting by its own dynamic intensity— the elemental vibrations of Nature—arouses these spiritual forces to the extent of responding to your call. When this can be done, but not until then, will your magical incantations have any effect upon the voiceless air. Not the priestly robes nor magic sword, not the incantations, WRITTEN WORD, nor mystic circle, can produce Nature's response to Occult rite; but the fire of the inward spirit, the mental realization of each word and mystic sign, combined with the conscious knowledge of your own Deific powers—this, and this only, creates Nature's true magician.

Who and where can such be found? Are they so few that the echo answers back "Where and who?" Yet, there are many such upon the Earth at the present time, but the present mental conditions forbid them making their identity known. They would not be recognized and accepted as the TRUE teachers, but reviled and persecuted and dubbed as insane. But silently, they are sowing the seed of truth that will spring up and bear fruit, where and when least expected.

Because evil is so active, truth is not lying dormant. The spirit of God, that Divine spark of Deity within every human soul, never sleeps, never rests. "On and upward" is its cry. "Omnia vincit veritas."

The grand sublimity of man's conception of at-one with the Infinite Father, at-one with the limitless universe of being, at-one with, and inheriting, all the sacred rights and inalienable prerogatives of the ineffable Adonai of the deathless soul, is the only test of man's qualification for the holy office; for, as Bulwer Lytton has truthfully said, "the loving throb of one great HUMAN HEART will baffle more fiends than all the magicians' lore." So it is with the sacred ritual. One single aspirational thought, clearly defined, outweighs all the priestly trappings that the world has ever seen.

The success of all incarnations depends upon the complete unison of VOICE and MIND, the interblend of which, produces the dynamic intonation, that chords with the inward rhythmic vibrations of the soul. Once this magical, dynamic, vibration is produced, there immediately springs into being the whole elemental world belonging thereto, by correspondence. Vocalists who hold their audiences spellbound do so by virtue of the magical vibrations they produce, and are in reality practical, even though unconscious, magicians. The same power, to a degree, lies in the voice when speaking, the graceful movement of the hand when obeying the will, and the eye rays forth the same dynamic power and becomes magical in its effects.

These powers are exercised more upon the physical plane, and no better illustration can be given, than, the power man is able to exert over the animal when gazing into its eyes.

Here, as well as in incantations and invocations, within the power of the will, lies the success or failure.

At this point it may be asked, what, then, is the use of magical rites, of symbols and priestly robes? We answer, in themselves alone, nothing, absolutely nothing, except the facility and convenience we derive from system, order and a code of procedure. To this may be added the mental force and enthusiasm of soul which such things inspire, just as men and women may feel more dignified, artistic, and refined, when dressed in accordance with their ideas. So may the average priest feel more priestly, holy; and consequently, more powerful mentally; when arrayed in the robes of his office and surrounded by the outward symbols of his power and functions. But, in themselves alone, there is not, nor can there be, any real virtue. The same may be said of the incantations. The words used in their composition are the hieroglyphics of mystical ideas. Therefore, the correct pronunciation of the words or the grammatical construction of a sentence is nothing, if the underlying idea is conceived in the mind and responded to by the soul. Will and motive form the basis of true magic.

One word more and we have completed our subject. Magic swords, rings, pentacles, and wands, may, and often are powerful magical agents in the hands of the magician, by virtue of the power, or charm, that is invested within them when properly prepared; but apart from such preparation, by those who know, they are as powerless as unintelligible incantations.

All the foregoing are aids, but if physical manifestations of magical forces be required, there must always be present the necessary vital, magnetic pabulum, by means of which such phenomena are made to transpire; and in every case, to be successful, the assistance of a good natural magician, or seer, is necessary; for without this essential element the whole art, in its higher aspects, becomes abortive.

CHAPTER XI. THE MAGIC WAND

This is the last lesson of our present course that requires a clear definition of the terms employed in the title thereof, for the twelfth, and final study is, perhaps, fortunate in having for its title a word that has not, so far, been misused and distorted from its original sense.

The Magic Wand. The words savor of everything that the young tyro in Occult art can picture to his mind; of the midnight magician and his mysterious, if not diabolical, arts, muttering his incantations, working his gruesome spells, and raising the restless ghosts of the dead. Strange fancies, these, and yet, so corrupt and ignorant have become the conceptions of the popular mind regarding the once sacred Science of the Temple and the psychological powers of Nature, that we very much question, if the ideas above stated were not very similar to the originals of each modern student, before he had become acquainted with the deeper truths—the realities of Occult philosophy.

We will commence our study by a careful investigation of the original meaning of the words Magic Wand, since those who were the masters and originators thereof, are far more likely to know more about them than their degenerate offspring of a later age. Few, comparatively, would believe that the words MAGIC, MASON, and IMAGINATION, are the present unrelated descendants of the same original conception—THE ROOT IDEA; but such is the case. First, then, we will examine their modern meanings. Magic is the unholy art of working secret spells, of using invisible powers, and holding intercourse with the unseen world of ghosts and demons, by means of enchantments. It also means the expert deception of the senses by the tricks of a conjurer, SO-CALLED hocus-pocus and fraud, and a magician is either an evil-minded, superstitious mortal, fool enough to believe in charms, or an expert pretender and imposter of the first water, who cheats and deceives the people. A mason is the honorable designation of a builder, who works in stone; metaphysically, a member of a semi-secret society, whose sole advantage is social intercourse and standing; who proclaim fraternity and universal brotherhood theoretically and practice the reverse in reality; a man who apes the Egyptian Mason, knows nothing in reality of Hiram, his master; who knows nothing of the starry Solomon or his mystic temple in the heavens, which Hiram built; and who misconceives the import of the three villains, or assassins; and who, further, knows nothing of that wonderful sprig of myrtle:—in short, a Free Mason, speaking generally, is a man who delights in ideals, social equality, secret fraternity, and plays at mysticism; who parades on the Masonic stage and enacts a role he does not understand. The first meaning, that of a builder, is the most correct. Lastly, the imagination is the exercise of mental imagery—the picturings of silent thought.

And now we will proceed BACKWARDS. Imagination is from the word "image," a form, a picture, and has descended to us from the Latin "imago," which, in its turn, was derived from the old Semitic root, "mag." Mason comes to us from the Latin "mass," which means to mould and form, i.e., to build; and the word "mass," through various transformations, was also derived from the root-word "mag." Consequently, originally, there was but little difference in the ancient idea of building pictures in the mind and erecting the mental idea externally in stone. It is from this fact, that, we have to-day Mental Masons, a la the secret orders, and stone masons, who labor for wages. The Mental Masons have merely lost the knowledge of their art. They should, by rights, be as active and correspondingly useful to-day as their more physical brothers, the masons of stone.

This art would never have fallen into disgrace and disuse, if their daily bread, or material accumulations, had depended upon their efforts in building up the mental, moral, and spiritual attainments, of each other, and bringing their knowledge into more external use, by making the material edifice, the physical body, a purer and more fitting temple, for the Divine soul.

Magic comes from the Latin "magi" and the Greek word "magos," which means wise, learned in the mysteries, and was the synonym of wisdom. The initiated philosopher, the priest, and the wise men, are all of them included in the "magi." Again, tracing this word to its remote ancestor, we find it terminating in the same Semitic root, "mag," but of this strange root no one was able to say much, except that it seemed to belong to the Assyrian branch of the great Semitic race. But quite recently, thanks to our scientific explorers and archaeologists, versed in the mysterious meaning of cuniform inscription; Assyrian scholars now inform us that they have found the hoary, primitive original of it, of magic, magi and imago, etc. It is from an old Akkadian word, "imga," meaning wise, holy, and learned, and was used as the distinguishing title of their wisest sages, priests, and philosophers, who, as may be supposed, gradually formed a peculiar caste, which merged into the ruling priestly order. The Semites, who succeeded the old Akkadian race in the valley of the Euphrates, as a mere matter of verbal convenience, transformed many of the old Akkadian words to suit their own articulation, and "imga" became "mag," and thus "magi." THE BLEND between the Semetic and the older Akkadian race, produced, by fusion of racial blood, the famed Chaldeans. So that we see how old are the words which many of us daily use, but with different meaning. Verily, it makes one feel, when be thinks of magic and its origin, as though he were quite nearly related to the people who honored King Sargon, the Wise, the earthly original of the mystic Solomon of Biblical tradition. The term Wand is an old Saxon word, which primarily signifies to set in motion, to move. From this we derive our word wander, i.e., to roam, and wandering, i.e., moving and continually restless.

We have now the original, therefore real, meaning of the words Magic Wand; thus an object that sets in motion the powers of the magician, and the magician, an Initiate of the sacred rites—A MASTER OF WISDOM, possessing all the resources that enable him TO BUILD mould, and form; to create in fact, by virtue of his knowledge of the secret powers of mental imagery and the potential use of his own imagination. He is both Mental Mason and learned philosopher.

The student may doubtless ask, why all this care and labor regarding mere definitions? We reply that, it is because, the real meaning of the words we have purposely selected for the title of our studies are, in themselves, a far better revelation than we could possibly have written. Originally, ideas and words were related as absolute expressions or correspondences, of each other. This is not so now. As the different races became interblended, the purity of both language and morals retrograded, and the people grew more to the external. The intuitions and spirit were compelled to retreat, giving place to only the intellectual and mental. The blending of the languages gave birth to many words wherein different meanings were transmitted; hence, the trouble arising to-day over the numerous interpretations of a single word.

Hybrid races have no such thing as a pure language. Their ideas and language, like their blood, is badly mixed up, confusing, and unsatisfactory, so far as the real meaning of the words are concerned. For this very reason we find so many different meanings for the same word; and also for this reason, we cannot formulate a legal enactment in the Anglo-Saxon tongue that, a learned lawyer, versed in this senseless jugglery of words, cannot demonstrate, to the satisfaction of the courts, means something the very opposite of the real intentions— the spirit—which the framers thereof, intended it to convey. Anciently, it required no artful cunning of the lawyer to interpret the laws. The words had only one simple and obvious meaning. If a language could be so constructed to-day, and the antiquated precedents of the courts annihilated; the legal profession would be exterminated inside of twelve months, and an affliction removed from the people.

The philosophy of the Magic Wand is this. It is a magnetic, electric conductor for the magician's will. It directs the flow of his thought and concentrates it upon a given point in space or an object. It is, magically, what the sights of a rifle are to a sportsman. It enables him to focus his powers with exact precision upon the mark against which, or upon which, his will is directed. Apart from this there is no power, per se, in the Wand itself, any more than there is in a lightning conductor without the electric storm. Ergo, the Wand is the conductor, in the magician's hand, for the lightnings of the soul; and just as the lightning rod is most useful and most powerful to protect, when the storm is the strongest; so is the Wand most powerful in the hands of the most potential magician. We can only transmit through this Wand the degree of force we may happen to possess in the soul.

In a properly prepared Wand lies the most powerful weapon, to protect or destroy, that can be placed within a magician's hands. With his own spiritual force and knowledge, combined with the magic power attached to the instrument, nothing can withstand its power, when directed with a determined and powerful will.

Many substances have been employed in the manufacture of these Magic Wands. Metals or stones will not serve this purpose, unless covered with some organic matter. In any case stones are worthless. The very finest Wands are made from the live ivory of a female elephant. A short Wand, twenty-one inches long, tipped with gold at the largest end and silver or copper at the other, is very powerful. Next to these costly articles are Wands with a gold or copper core, a wire, in fact, cased with ebony, boxwood, rosewood, cedar or sandalwood. English yew also serves the purpose; so does almond wood. Simpler, less expensive, and almost as effective, are Wands made of witch-hazel. In fact, apart from the Wands of live ivory, I consider that witch-hazel is as powerful as the golden Wand. Next in force to this witch-hazel are the shoots of the almond tree, and, lastly, the peach and swamp willow.

The proper time to manufacture a Magic Wand is whenever you can find the person who is able to do the work. But after it is constructed it must be thoroughly magnetized, with proper ceremony and aspiration, the first or the second full Moon after the Sun enters Capricorn, at midnight, when the Moon will be culminating in her own sign upon the mid-heaven.

The best time TO CUT a shoot of witch-hazel or other material for a Wand is the first full Moon after the Sun's entry into Capricorn, at midnight, and then magnetize it upon the next full Moon at the same hour.

In conclusion, let us repeat that, the Magic Wand is but the highly sensitive magical medium for transmitting and concentrating the force of the learned magician; that it is equally powerful under great excitement of mind, WHETHER USED CONSCIOUSLY OR NOT. The stream of mental fire will go in the direction the Wand happens to be pointed, and, therefore, should never be in the hands of the wicked or foolish, any more than firearms. It is potential or otherwise, in exact proportion to the artist's wisdom and dynamic mentality, and is useless in the hands of the idiotic or weak-minded. A Magic Wand requires brains and vigorous mental force to make it effective, just as the steam engine requires an apparatus for generating the steam, that moves it. With a determined will, and a mental conception of one's inward power, any man or woman can, by means of this sensitive Wand, defy all the legionaries of Hell, and quickly disperse every form of spiritual iniquity.

The firearms which have become so intricate in their mechanism and so destructive in their operations, are only a degeneration of the Magic Wand. The first weapons of warfare and slaughter were very crude and clumsy, then larger and more destructive, until at last they have become as fine in texture and mechanical genius, compared with their early brothers, as the Magic Wand is to-day, above and beyond, the present weapons of warfare. At last, the original mode of defense will be rediscovered and become a utility in the hands of the majority of mankind. At the same time, the mental and moral nature will be evolving into better conditions, too, so that their use will not be given to the ignorant and evildoers, but placed in charge of the educated, those who are morally capable of leading and ruling.

Yes, we are now stepping upon the plane of reason and intuition, where right, not might, will prevail and rule the world. The present mode of government and rule will be changed, and one of humanitarian justice take its place.

God hasten the Millenium.

THE BOOK WHICH IS CALLED THE TABLETS OF AETH

THE SACRED SCROLL WHICH IS CALLED THE TABLETS OF AETH

NOW FOR THE FIRST TIME TRANSCRIBED FROM THE ASTRAL RECORDS AND DONE INTO A BOOK,

By ZANONI

TO WHICH IS ADDED A SERIES OF INTERPRETATIVE REFLECTIONS FOR THE SPIRITUAL MEDITATION OF THE FAITHFUL.

FOREWORD

> Thy temple is the arch Of yon unmeasured sky; Thy Sabbath the stupendous march Of grand eternity.

To my Brothers and Sisters of the Hermetic Brotherhood of Luxor:

GREETING—For some years it has been my desire to leave a spiritual legacy to the many devoted friends and followers who have braved so much amid present truth and error for my sake.

In choosing the present work for such a purpose, I have had in view the deeper spiritual needs of the soul—the prophetic element of the interior spirit, which can best exalt itself through the contemplation of Nature's arcane symbolism of the starry heavens—not the material expression of the glittering splendors of the midnight sky, but the spiritual soul-pictures of those blazing systems that reveal to the seeing eye the shining thrones of THE RULERS—the Powers that Be.

Ever since the dawn of intellectual human life upon our Mother Earth, long before the days of the cave man, or even the first frost that heralded the coming of the Ice Age, souls have hoped and hungered and souls have quailed and fallen in their struggles with the mysteries of God. But ever and anon some bright flower of the race has gained the spiritual victory. A Messianic soul has responded to aspirations of a great-hearted, great-souled woman, pregnant with spiritual yearnings beyond her race, and she has unconsciously blessed her kind for the generations yet to come with that incarnated mystery—THE SON OF GOD. Blessed, O Woman, is thy patient mission on the earth, and transcendent are the holy mysteries of thy maternity. Every human birth is a Divine miracle in humanity, performed by the Motherhood of God.

Hence it is that, from the earliest ages of life, triumphant souls have stormed the gates of the sanctuary and penetrated Nature's most occult mysteries and there recorded their spiritual victories. Amid these sacred records lies one great scroll, that none but the brightest and bravest may read.

This sacred scroll, sealed with the seven mystic seals of the heavens, contains The Tablets of Aeth, a record of the soul's experiences upon the planes of both conscious and sub-conscious life-spirit and matter, that are expressed in a series of universal symbols, which manifest to the seer the processes of creative life, of spiritual cause with material effect. And, finally, the mystery of the seven vials and the seven stars of Saint John are written therein; for the Tablets are the hieroglyphic keys which unlock the realities of truth involved within the unrealities of external life, and open up, to the aspiring soul, inconceivable vistas of knowledge yet possible of realization, within the Divine womb of the uncreated Aether.

Myriads of exalted spirits, who have toiled for the treasure which doth not corrupt, have added, and are adding, their portion of personal conception to this universal conception of life, so that the sacred symbols themselves, inscribed upon these imperishable Tablets, ARE EVOLUTIONARY—are slowly unfolding through the eons of time, and revealing wider and yet deeper processes of the light, life, and love, of the Motherhood of God.

Therefore, all Divine revelation of infinite truth is limited and finite as to its conception, when revealed through a finite capacity. All Divine truths are universal; all personal conceptions of such truths are limited; hence springs the unquenchable fountain of the ONE eternal truth, eternally repeating itself, in cosmic as in human life, by the progressive unfoldment of Nature's unlimited potentialities.

"The outward doth from the inward roll,
And the inward dwells in the inmost soul."

The true poet is always a seer, and he might have added that the INMOST SOUL is the uncreate, and, the yet uncreated itself, lies buried in the ever eternal beyond; hence the immortality of the human spirit.

This sacred astral scroll, rightly and reverently studied by the disciple of the higher law, becomes a boundless source of knowledge and inspiration. There is no mood of the mind or yearning of the soul that cannot be satisfied and refreshed from this inexhaustible fountain of spiritual truth, no passion of the human heart that cannot be eased of its burden and soothed of its pain. Its spiritual refreshment falls like the dew from heaven upon those who are weary and heavy laden with the trials and sufferings of external life.

Accept it, then, even as it is given unto you. My friends and brethren, accept it as Zanoni's last work on earth—his legacy to you, and may the spirit of the All-Father—

Mother, the ineffable spirit of Life, Light, and Love,—the Unknowable, whom men call God, rest upon you and be with you now and forever.

INTRODUCTION

TO THE BOOK WHICH IS CALLED "THE TABLETS OF AETH," WHEREIN ARE DESCRIBED THE FORMULAS OF MEDITATION.

THE FORMULAS OF MEDITATION,

TO THE DRAGON, FOUNTAIN OF YOUTH.

"When first, a musing boy, I stood beside
Thy starlit shimmer, and asked my restless heart
What secrets Nature to the herd denied,
But might to earnest hierophant impart;
When lo! beside me, around and o'er,
Thought whispered, 'Arise, O seeker, and explore.' "

The Tablets of Aeth are the culminating expression of symbolical ideas, and the studious meditation thereof is to be approached and continued in this wise:

First, commit to memory, as near as may be, all the ideas involved in the astrological laws and principles laid down in "The Science of the Stars," formulated in the second part of "The Light of Egypt," Vol. I, especially as regards the symbolism there given and manifestation thereof on the intellectual plane. Mentally digest these aspects of truth most thoroughly.

Second, carry forward the same course of mental training with regard to the preceding chapters in this volume, from No. 1 to No. 12. There are thirteen chapters, but No. 13, the last one, being "The Penetralia," should not be included in this course, but, rightly used, should be reserved as the last and final revelation for spiritual contemplation.

The twelve chapters just mentioned continue the great astral laws given in "The Light of Egypt," Vol. I, from this plane to that of the soul life of the human monad (both prior to and after human incarnation). At this point we leave the finite and step into the realms of the infinite. From the sphere of limitations which surround the microcosm we enter the starlit path of the macrocosm, and here, with the illimitable ocean of eternal life sweeping onward before us, we hear the first strains of the Grand March of the Universe burst forth from the organs of God! The suns of creative life swell the infinite chorus of sound; archangels swing their fiery batons to the march of the heavenly host; and all earthly sound has ceased. We are absorbed in the music of the spheres.

We are now in the realm of universals, the domain of living realities. The Tarot of Mother Nature revolves before us, revealing her mystic meanings to the soul. All ideas are symbols, and symbols are reservoirs for the conservation of thought. And this is a very truth: Even so on earth as it is in heaven.

The Tablets of Aeth, then, constitute a spiritual astrology, a spiritual science of the stars, void of mathematics, yet possessing all the exactitude of figures, constructed on the principles of astronomy, yet expressed by the methods of the Kabbalah.

The transmission of spiritual truth from inward to outward form, though differing according to the age in which it is expressed, is ever the same in principle. And in the same way that the sacred clavicula of Solomon became the Tarot of Bohemian gypsies, so did the Tablets of Aeth manifest their mysteries in the starry science of Chaldean lore. But there is this sharp line of demarcation between them, namely, the Tablets of Aeth deal with universal human life and nature, with infinite principles from which all finite laws radiate. The Tablets of Aeth express and symbolize the cause. All other mundane systems of occult study, astronomical or metaphysical, are spirito-natural effects, the individual intellectual fruits, gathered from the one universal tree of knowledge. Uncreated, Unlimited Potentiality, is the one impersonal truth shining forever in the Great White Light of God. All the laws, powers, and principalities, manifested in the moving Universe, are but the colored rays, blazing with glorious life through the prisms of matter.

Having stated thus much, the neophyte will perceive in what meditative sphere of thought the Tablets may be used. The method of study is, as shown, a purely synthetic deduction of human ideas from spiritual symbols of universal principles. The Tablets themselves constitute a grand arcane Tarot of man, God and the universe, and of all the powers that dwell therein. They may be studied singly, as, for instance, meditating upon some one great universal idea or principle; or they may be studied in trines, as they appear in each separate book, or chapter, or as squares, like two, five, eight, eleven, or as the seal of two trines, one, three, five, seven, nine, eleven, with No. twelve in the center, as the revealer of the mystery. And, finally, they may be contemplated as the Grand Oracle of Heaven, in the following manner:

Make a circle of the tablets, as you would with a pack of Tarot cards, beginning with No. 1, on the eastern horizon, and proceeding in the exact opposite order from a figure of the heavens—No. 2, being on the Twelfth House, No. 3, on the Eleventh, and {} on the M. C. of the figure, as in the Astro-Masonic chart, given in the second part of "The Light of Egypt," Vol. I, and so proceed with the rest of the twelve tablets of the stars. This figure will represent the potentialities of the macrocosm, the starry signs symbolizing the possibilities of things past or to be, and the rulers the active executors thereof. Study the figure in all its aspects as such, first singly, tablet by tablet, then as a whole—the cosmos. Next, place the ruler of any given tablet at the side of the Mansion, and try to penetrate its various meanings, powers and possibilities. Then proceed the same with a trine and a square, and, last, with all the rulers, in the order of their celestial lordship of the signs, each in his appointed place, as a whole Arcana.

In any grave crisis of mental or physical affairs, wherein nations, and not individuals, are concerned, the tablets may be used as a celestial scheme of the heavens, thus: Cast a figure of the heavens for the Sun's first entry into the sign Aries at the vernal equinox, calculated for the meridian of the capital city of the country under consideration. Degrees and minutes are not wanted. Then place the twelve tablets in place of signs, exactly as they would occur in an astrological figure. Then place the rulers of the Sun, Moon and planets therein (each having its own tablet), as they are found to be situated in an ephemeris for the time of the figure. This done, study the whole from a spiritual standpoint as the causes and ultimates of the crisis, according to astro laws.

The foregoing simple directions will, I think, be sufficiently plain for all purposes, never forgetting that this holy study is not a system of divination, as commonly understood, but of Divine revelation, in its highest and most holy religious sense. Long study and most reverent meditation will be required to master this mystery, and many errors of judgment will occur to the beginner.

The interpretative reflections are added for the purpose of guiding and guarding the spiritually untrained seer from possible error in fundamental conceptions only. They must not by any means be taken as a complete revelation of the tablets, but only as a series of skeleton keys by means of which all things may be revealed to the earnest seeker thereof. To have added more than is given would only be to defeat the object of this work. Each seeker for the truth must excavate the mines of knowledge, and dig further into this universal well of truth for himself.

Remember that all interpretation will be personal to each student. Of no one can it be affirmed, "thou hast said," and so endeth the matter. Not so. To each, according to his talent, shall the mysteries of the kingdom be revealed, to every one according to his humility, spiritual light, and merit. But from the arrogant, the selfish, and spiritually proud, shall all things be taken away, and truth shroud herself in the veil of delusion. In simplicity of mind, then, and purity of soul, approach the Holy of Holies. "Suffer little children to come unto Me," saith a messenger of the Most High, "for of such is the Kingdom of Heaven." Verily, therefore, I say unto you, that not until you can look upon all the works of Nature—beauty in her nakedness or vice and crime in their repulsiveness, with pure thought and holy feeling, can you inherit eternal life.

Here endeth the introduction to the book which is called "The
Tablets of Aeth."

PART I

OF THE TWELVE MANSIONS

Here beginneth Chapter I of the Book which is called "The Tablets of Aeth," wherein is transcribed the First Quadrant of the Twelve Mansions.

"I sent my soul through the invisible,
Some lesson of that after life to spell;
And by and by my soul returned to me
And answered, 'I, myself, am Heaven and Hell.' "

"The moving finger writes; and, having writ,
Moves on; nor all your piety nor wit
Shall lure it back to cancel half a line
, Nor all your tears wash out a word of it."

TABLET THE FIRST

Aries

SYMBOL

A deep blue Sky, a blaze, as if something were about to rise.

I

REFLECTION

TABLET THE FIRST

The blush of dawn of a new life, all nature quivering with the sense of coming, conscious life; Isis, vibrant with love of the coming child, her bosom flushed in expectation of the little son soon to breathe on her yearning breast.

In this we trace the great lesson of preparation, of sending the light before the form, of the prophecy before the fulfillment. Dawn must precede sunrise. What you expect will be your destiny.

It is the longing of centuries that incarnates a god, a real Sun-God, whose vibrant love-life can thrill other lives into prayer—aspiration, the struggle for eternal life. The dawn represents the expectant maternity of Nature—God.

O child of Adam! See that thou expecteth much, and that thy aspirations are reflected in thy outward life.

TABLET THE SECOND

Taurus

SYMBOL

A red sun on the horizon of an inky sea.

II

REFLECTION

TABLET THE SECOND

Nature has shown forth her glory, as brought forth by young Horus, but her creative force is still unreflected. The sea is black and inky. The Son of God is born, but the sea of human life still remains unconscious, in primeval darkness.

The angles of the Sun and the sea are not yet in right relation to each other. A few, standing on the watch-towers of life, seeing the red glow of the risen sun, call "Look!" But the unfortunate ones in the outer darkness cry, as they beat their breasts; "No! There is no light! You do but dream!" And yet the Sun of Life has risen—the Divine light glows.

O child of Adam! Remember that "In Him was life, and life was the light of man, and the light shineth in the darkness, and the darkness comprehended it not."

TABLET THE THIRD

Gemini

SYMBOL

Two stars are rising at angles to each other and to the Polar star, while eight stars shine faintly in the black space of background.

III

REFLECTION

TABLET THE THIRD

The Divine symbol of soul-matehood is here signified in the two stars rising in the foreground; not only the soul-affinities of humanity, but the eternal father-mother forces manifested in the biune spirit of universal life and nature, the two great creative powers, Life and Light, whose harmony creates love, attraction and repulsion, and the straight lines of law and justice, which blend in the spiral of mercy.

The two stars are rising at an oblique angle to the pole-star, the center around which, material things revolve. So, too, life and love are balanced by the star of wisdom. Love in the spirit is adaption to the environment in matter and providence in universal life. The eight stars reveal the mystery of the tablet—universal death, present with life, the final end of all discord glimmers faintly afar off, and man questions the love of God, seeing that all things pass away, not realizing that death is the germinal promise of life, of transformation, of the realization of unrealized hopes, of the union of loving hearts in their starry pilgrimage back to the Father's home.

O child of Adam! Listen unto the words of the Teacher: "I and the Father are one." Suffer little children to come unto me, for of such ii the Kingdom of Heaven."

PART I

Here beginneth Chapter 2 of the Book which is called "The Tablets of Aeth," wherein is transcribed the Second Quadrant of the Twelve Mansions.

"How they struggle in the immense Universe!
How they whirl and seek!
Innumerable souls, that all spring forth
From the vast world-soul.
They drop from planet to planet,
And in the abyss they weep
For their forgotten land.
These are thy tears, O Dionysus,
O Spirit vast, Divine One, Liberator.
Draw back thy daughters to the breast of light."

"Ah, love! Could you and I with him conspire
To grasp this sorry scheme of things entire,
Would we not shatter it to bits? And then
Remould it nearer to the heart's desire."

TABLET THE FOURTH Cancer

SYMBOL

A woman's face unconscious, in trance, surrounded by clouds.

IV

REFLECTION

TABLET THE FOURTH

The dreaming woman, whose brooding thoughts shape the coming man. The race is never any farther advanced than the average thought of the woman. She is yet sleeping, knowing not her powers. So, not until she awakes and recognizes herself as conceiving by the Holy Ghost and the mother of the incarnate God, will that God be brought forth unto universal knowledge.

In this is the great lesson to woman: Ever remember thy creative power as the mother of the humanity of the future. The sun in thy mansion exerts its highest power. Awake, therefore, O soul, and eclipse not its brightness with thy dreams of sublunary power.

O child of Adam! Ever honor the womb that gave thee birth, and know that all thy earthly greatness received its seed therefrom. A fountain cannot rise higher than its source.

TABLET THE FIFTH

Leo

SYMBOL

A man's arm, bent, exceedingly muscular, a knife in the hand, a streak of lightning opposite the arm, which is defying the lightning.

V

REFLECTION

TABLET THE FIFTH

Here we have the symbol of the incarnate fire of the spirit defying the mere natural fire of the heavens. The woman sleeps and broods and dreams, but the man she has brought forth is awake, and bids defiance to the fiery forces of Nature. He has armed himself with the keen knife of action, and with it has conquered the forces of matter. He has harnessed the lightning, and made the electric fluid his obedient slave. And thus has he mastered all forces inferior to spirit—that spirit of conscious life which is his birthright.

The lesson to be gleaned from this is that, the kingdom of Nature must be taken by storm. Not for rest, but for work, has Mother Nature sent forth her man child; not for peace, but for battle; not for inertia, but for effort.

O child of Adam! Arm yourself with the sword—mayhap the sword of affliction—and, gallantly raising the strong right ann aloft, hurl defiance at the chaos of Nature, sure that the fire from the Sun of the spirit is burning in every vein of that arm.

TABLET THE SIXTH

Virgo

SYMBOL

A Lotus, rising from the water, coiled around its stem a snake, whose efforts fail to reach the flower.

VI

REFLECTION

TABLET THE SIXTH

Here we have the sacred flower, symbol of the virgin soul, uncontaminated by the snake of passion, which can only enfold the body— the stem; the snake of matter—of lust—of evil. But the flower of the spirit—the soul—lifts its pure white petals upward as an incense cup to the Sun of the Spirit.

In this symbol read the great lesson of the experience of evil. If, the flower of the soul, blossoms; the mud of the soil and the snake of the passions are but the surroundings of its roots and stem. Both are necessary for the perfection of the flower. The roots sink deep into Mother Earth, and draw nourishment and life, lifting matter upward, while the snake of passion becomes, under another aspect, the serpent of wisdom. Coiled around the stem of this life, it gives to the incarnated soul that wisdom which later blossoms in the Seraph of the Sun spheres.

O child of Adam! Take suffering, if it forge the sword of the spirit. Take evil and passion, and turn them into deep lessons of life, blossoming the evil into good, changing passion into wisdom. Only "the pure in heart can see God."

PART I

Here beginneth Chapter 3 of the Book which is called "The Tablets of Aeth," wherein is transcribed the Third Quadrant of the Twelve Mansions.

"To know what really exists, one must cultivate silence with ones self, for it is in silence that the eternal and unexpected flowers open, which change their form and color according to the soul in which they grow. Souls are weighed in silence, as gold and silver are weighed in pure water."

"The worldly hope men set their hearts upon turns to ashes; or it prospers, and anon, like snow upon the desert's dusty face, lighting a little hour or two, is gone."

TABLET THE SEVENTH

Libra

SYMBOL

A crowned king, with a scythe raised in the air, looks closely at two boys wrestling beneath him in a field of grain, a red poppy below them.

VII

REFLECTION

TABLET THE SEVENTH

The symbol of Nature's eternal war for the impossible equilibrium between spirit and matter; the symbol, also, of Time, which is but the illusion in which eternity clothes itself; forever putting on and forever putting off new garments of matter. The crowned king is the victorious soul, waiting, with the scythe of Time, to reap the harvest of the world; while incarnated man, as represented in the wrestling youths, is struggling for that which he did not produce, and which only death can reap. The poppy reveals the secret of the illusions of Nature's master-showman. All earthly things are unreal to the spirit, which is the only real thing. Man's effort to hoard and save the things of this world IS INJUSTICE TO OTHERS. The struggle is eternal, and no matter how careful or cunning man is to monopolize either power, truth or wealth, swift-footed time will readjust all things without error.

O child of Adam! "Lay not up for yourselves treasures upon earth, where moth and rust doth corrupt and where thieves break through and steal."

TABLET THE EIGHTH

Scorpio

SYMBOL

A wide, and plain, on it a skeleton; a dull, grey sky, in which an Eagle soars, full-fed, it seems, from the flesh of the skeleton.

VIII

REFLECTION

TABLET THE EIGHTH

A significant symbol to the seer, showing forth the two ultimates of life and death, of earthly things and sex. Scorpio is both the eagle of the spirit, soaring aloft, well fed with all that is worth carrying away from the earth; and also the scorpion, whose natural home is the desert.

In sex, either way, life is given. Shall it be to your spirit making fat and full your immortal self, or will the other interpretation be yours? And will you leave yourself dead and annihilated, a skeleton, to the Ego, the Divine spirit? For sex is indeed the foundation of all. Raised to the region of Libra, it is power and magnetism. To the bosom it is love; to the brain it is enthusiasm. It is the promethian fire of life, the creative force, giving vigor to whatever region to which it is raised; or, lowered, to be spent with no returns, it debases and renders life a desert of dry bones.

O child of Adam! Reflect on the fall of man from spirit to matter, and combine the wisdom of the serpent with the purity of the dove, and "lest ye partake of the tree of life ye shall surely die."

TABLET THE NINTH

Saggitarius

SYMBOL

A child in a shell, holding in its hand a feathered lance, is drawn by five stars, grouped in an under arc.

IX

REFLECTION

TABLET THE NINTH

The symbol of the conscious soul. The shell is the body, drawn by the five senses—stars—which form an under arc, to represent the world of material things and our relation thereto. The child, armed with the feathered lance, is the soul; riding thus, fully armed, in the shell of the body, it realizes the duality of truth; that all things are changeable; and that each thing is true upon the plane of its manifestation, while an illusion to that which is interior to its life, while the soul is in its dream state. Sagittarius represents conservatism and the permanence of crystallized institutions; but, when the spirit awakes and bursts the shell of matter, the senses, instead of being the guardians and jailors of its environment, become its servants, and the means by which, united as the one Ego, sense-perception, it races o'er the fields of Aeth—a being of life and beauty, shining in the empyrean of God.

O child of Adam! Ever remember that temperament and environment constitute the north and south poles of human possibility, and that ability, combined with opportunity, is the measure of responsibility.

PART 1

Here beginneth Chapter 4 of the Book which is called— "The Tablets of Aeth," wherein is transcribed the Fourth Quadrant of the Twelve Mansions.

"A hair, perhaps, divides the false and true. Yes, and a single alif were the clue— Could you but find it—to the treasure house, And, peradventure, to THE MASTER, too.

"Beware, O my son, of self-incense. It is the most dangerous on account of its agreeable intoxication. * * * Learn, O my beloved, that the light of Allah's truth will often penetrate an empty head more easily than one too crammed with learning."

TABLET THE TENTH

Capricorn

A deep, black ground, o'er which shimmers a phosphorescent light; at each side an aurora borealis rises, mountain like; above all, a tiny star.

X

REFLECTION

TABLET THE TENTH

Here is revealed the symbol of the messenger of the Most High. The star hovers over the phosphorescent light cast on the darkness as the spirit hovers over the blackness of matter. The aurora borealis stands as the emblem for the magnetic attraction of Earth on spirit, the Christ soon to be born in the manger of the Goat; the descent of the Holy Ghost into material form, so that heavenly truth may illumine the drear speculum of earthly thought with the Divine iridescence of celestial light. It is the lowest arc of the cycle that reveals the new birth of death unto life—the divine egg of Brahma, containing the promise of the new law: "Peace on Earth, good will towards men."

O child of Adam! Be thou the star, and not a dweller of the outer darkness, and "Let your light so shine before men, that, they may see your good works."

TABLET THE ELEVENTH

Aquarius

SYMBOL

A stormy sea is seen; above it the eight stars shine, brilliant and clear.

XI

REFLECTION

TABLET THE ELEVENTH

This tablet symbolizes the complete materialization of man—man, perfect on the earth and the lord thereof, in so far as material forces are concerned. The storm is the tempest of life, the whirl of the elements of matter in their battle with the spirit. The eight stars, brilliant now (for they are the same stars that were dimly seen in Gemini), show that the conquest of matter is complete, the great fall of spirit finished; the end of involution. And this would bring stagnation and death, if peace now ensued. The lesson taught is that, not in peace and rest can the soul grow; but amidst the earthquakes that shake thrones, the floods that overwhelm countries, the fires that reduce to ashes, has the strong man-soul grown to its present state and power. So fear not the storm, but the calm; not the unrest, but the quiet; fear not the battle, but the ignoble peace of the coward.

O child of Adam! The astral soul must learn to do and dare. Not over the brave man's grave shall it be written, "Rest in peace," but "I will arise, and go to my father."

TABLET THE TWELFTH

Pisces

SYMBOL

A comet, beyond it infinite things, only dreamed of as yet, a world floating in an ocean and in night, beneath are two hands clasped palm to palm.

XII

TABLET THE TWELFTH

A REVELATION OF THE TO BE. The comet is the twelfth Avatar, the herald, coming forth from the starry abyss of the infinite, staying with us a little while, and then flashing on his shining way to other worlds than ours, bearing THE DIVINE WORD from sun to planet, as the fiery messenger of God. And here the soul may well ask: "Who? Where? Whence and Whither?" For behold, he has come, and gone, and

"Earth could not answer; nor the seas that mourn In flowing purple, of their Lord forlorn; Nor rolling Heaven, with all his signs revealed And hidden by the sleeve of night and morn."

The world floating in the sea of the infinite and resting in night shows the present state of humanity. But, "the blush of dawn" is ready to gladden the soul, and the expectant seer, from his lonely vigil on the hilltop, awaits the sunlight which will soon flood the world anew.

The two clasped hands point to many problems, chiefly soul- matehood, the message of the starry messenger, universal brotherhood, and the Father-Motherhood of God.

O child of Adam! Watch and pray, that a voice of the silence may speak unto you.

Here endeth the four Quadrants of the Tablets of the Twelve Mansions, wherein are revealed the signs and symbols thereof, as faithfully transcribed from the sacred roll in the astral records and called "The Tablets of Aeth." April, 1893.

PART II

of The Book which is called

THE TABLETS OF AETH OF THE TEN PLANETARY RULERS

PART II

Here beginneth Chapter I of the Second Part of the Book which is called "The Tablets of Aeth," wherein is transcribed the First Trinity of the Planetary Rulers.

"The human heart is the true temple of God; enter ye into your temples and illumine them with good thoughts. The sacred vessels, they are your hands and your eyes. Do I say that which is agreeable to God—doing good to your neighbors? But, first embellish wherein dwells He, who gave you life." ——

"How small soever your lamp be, never give away the oil which feeds it, but only the light and flame, which crown it."

TABLET THE FIRST

The Sun

SYMBOL

A flaming splendor, a center of light, radiating in all directions.

I

REFLECTION

TABLET THE FIRST

The symbol of all created life, spiritual and material; of all goodness, human or Divine; the center of all thought, from brutal instinct to Deific wisdom; of all creations, from starry systems to man, and from man back again to invisible gas; of all action, from the imperceptible vibrations of nerve energy to the awful destruction of worlds. All creative potency lies within a Sun sphere. Light is life. The planets are but the offspring of light and life. So in this symbol, we read the source of the human Ego, of our own life. We are, as it were, the planets of the spiritual Sun. Our souls are the attributes of the Sun, of the spiritual Ego. Only from the Ego can we receive life eternal and make immortality a fact. Obeying this spiritual life-force, the human monad is but an attribute, a reflection, of the Divine Ego, and if it fails to awake to a consciousness of this union, it withers and dies like a flower plucked from the parent tree of life.

O child of Adam, in reverence and awe do thou meditate upon this Tablet, for it is a thing of beauty, a being of light, life and love, manifesting its creative mission. It is the Vicegerent of God, flaming forth His splendors in the sky.

TABLET THE SECOND

Mercury

SYMBOL

An elephant, kneeling between two square columns; on one an eagle, on the other a vulture.

At the side a boy, with bow and arrows, standing in doubt which to shoot.

Below these a human face, composed of various flowers, whose roots are snakes, a poppy, forming an eye, which winks.

II

REFLECTION

TABLET THE SECOND

A vision revealing the earthly drama of the microcosm. The elephant represents the highest expression of intelligence, minus the spirit; kneeling between the square columns of matter, i.e., guarded by them. The external mind is sleeping, or, at most, dreaming of the things of the spirit. Above sleeping mind sit the two birds, who represent spirit and matter, each waiting for the slowly preparing feast. The boy, the soul with its weapons, has a choice. Shall it be the sensuality of the flesh that he shall destroy, or the possibilities of the spiritual life on earth. The problem awaits solution. The eagle sits ready to bear aloft the spirit of the sleeper. The vulture hopes for sleep to end in death, that he may live upon the carrion thereof. The flowers of the external mind have for their roots the snakes; and, in a larger sense, the flowers of immortality have the serpent of wisdom for their roots. And the poppy winks. It knows its own power of illusion, and the double significance of the snake; the necessity of evil in the evolution of good. It is the Tablet of Wisdom.

O child of Adam! "Be ye therefore wise as serpents and harmless as doves."

TABLET THE THIRD

Venus

SYMBOL

An altar: on it two cups, one full, the other spilled; near them two bleeding hearts, in one a snake, in the other a dagger.

Above—clouds, from which comes a woman's face, a wreath in the hand, coming out of the cloud; in the wreath an angel, going upwards, with wings outspread.

III

REFLECTION

TABLET THE THIRD

There is but one altar, but one blood of the sacrament in two cups, but one flesh of the Christ—the Ego—in two hearts, two experiences in love, ecstacy, and pain; two results of experience, the serpent and the dagger, symbolizing wisdom and affliction. Above the altar the divine woman holds the wreath encircling the angel. The angel of immortal life rises from the altar of sacrifice. Some of the wine is spilled as offering. The cup that is filled is raised to "Ra." To serve at the altar of love is the soul-mission of all, even as Christ served his disciples. Each soul must find its own service, and then the pilgrims of the Sun return to the mansions of the blessed. The great mother-god, Venus, Urania, quivers and thrills as she holds forth her offspring—the angel, the young Eros of life eternal.

O child of Adam, this is the Tablet of Love. Meditate thereon, as the last of the triune God. In this Tablet lies the secret of suffering and pleasure. He who vibrates in pain will quiver in ecstacy. Only those who have agonized in Hell can thrill in Heaven.

PART II

Here beginneth Chapter 2 of the Second Part of the Book which is called "The Tablets of Aeth," wherein is transcribed the Second Trinity of the Planetary Rulers.

"Thou art called forth to this fair sacrifice
 For a draught of milk; with the Maruts Come hither, O Agni!

They who know the great sky, the Visve Devas without guile; with those Maruts Come hither, O Agni!

They who are brilliant, of awful shape, Powerful, and devourers of foes; with the Maruts come hither, O Agni!

They who in heaven are enthroned as gods, In the light of the firmament; with the Maruts Come hither, O Agni!"

"Let us meditate on the adorable light of the Divine Rulers. May it guide our intellects."

TABLET THE FOURTH

The Moon

SYMBOL

NIGHT

A wonderful spider's-web;

The web glitters in the faint moonlight against a dark background of blue; moon invisible; on the outside of web a star, in the center a spot of light, underneath a coffin filled with stones.

IV

REFLECTION

TABLET THE FOURTH

The web of life has caught the monad of the soul and thus incarnated the universe, for each soul incarnates its universe at birth, each one's world being different, and peculiar unto himself. At the first breath, the young child polarizes his relations to stars and earth, and it is the affinity and repulsion which make his life experience. And the stars weave the web in their lines of sextile, square and trine, of opposition and conjunction, thus enveloping the monad in the Circle of Necessity.

Outside the star of the spirit, the Ego, shines clear, free from the entanglements of the web and unaffected by the magnetic glamour of the Moon. And lo! the coffin is filled with stones, a symbol of death and the Moon, which is but a casket of stones. Therefore, little monad, caught in the tangle of the web of life and the glamour of earthly things, take heart, for, beyond all, is the star of your being. Call down the law of that star into yourself, and the web is broken and waves its tattered shreds in the breeze. The moonlight, the reflected light, pales as the Star-Sun of your being rises, and the moonlight of Earth gives place to the Sun-spheres of Ra.

O child of Adam! The beginning of sorrow is the dawn of spiritual life. The wise man rules the stars; the fools of Earth obey.

TABLET THE FIFTH

Mars

SYMBOL

An immense helmet on pedestal, across which a streak of lightning flashes; beside it a naked child painting pictures on the helmet; beneath, a broken sword.

V

REFLECTION

TABLET THE FIFTH

Can greater irony be shown than in this astral symbol. Mars is externally represented as a fierce warrior, awful to behold; the reality, a little child, painting toy pictures on the helmet, too big for his curly head. The lesson in this is indeed, that the pen is mightier than the sword; that the big and blustering helmet will become a plaything for the child. Soon, that the sword of bloodshed, rape, and ruin, will be broken and war relegated to the past, looked at, but, as pictures, painted with hideous reality by the childhood of the race.

The symbol also reveals the great executive forces of humanity, the child. The soul can paint, execute its ideas, its hopes and its fears in any color—the lurid red of blood, the black of ignorance and crime, or in the living light of beauty. All the same, it is the childhood of man painting its ideals in the material world.

O child of Adam, curb the anger of Mars, that thy painting may set the dove at liberty. Let the magic of thy soul transform the savage of the desert into the angel of mercy.

TABLET THE SIXTH

Jupiter

SYMBOL

A cave in the mountain side; a face like the sphinx comes out of the cave, there is a blackness behind it; it looks with upturned head to a light that is way beyond; it is a face that means something awful, a godlike defiance to the things that are.

VI

REFLECTION

TABLET THE SIXTH

Again we are impressed with the contrast of internal and external things. Jupiter, the symbol of authority, conservatism, church, and state, and the stability of human institutions, and the things that are, as the things that are the best. But oh, how widely different the internal, the real Jupiter, that governing power of the spirit that hurls defiance at unjust authority, the cruelty and tyranny of the world. The soul sees the light beyond, and, emerging from the dark chasm of matter, knows the battle that must be fought against wrong. It is the awful—yea, terrible—symbol of defiance to gods and men who oppose its onward, upward march to the shining goal of light. Make way, then! Make way! For Earth has given birth to her giant son—the Spirit. For, listen closely, my friend, to the axiom of Immortality. What is soul? Not the spirit, mind you; not the deathless Ego, of which you at present, perchance, know absolutely nothing. Soul is mere memory; a scavenger in earthly states; and a gleaner, a hired help, in the fields of heaven; and to become immortal, there must be something more than soul as the result. It must take such a vital interest in its Lord's work that, finally it becomes too valuable to lose, and must be taken into partnership, so to say. The Ego—Lord— has found a valued servant, a trusted steward, after much seeking, and at once adopts it as its very own. And so the soul becomes heir to the heavenly estate and receives the immortal, vital principle of spiritual union, and awakes from the son of Earth a God-like being, free from the shackles of Time—a dweller in eternity. The soul must awake and realize the Deific atom around which it revolves before it is too late. Unless this is so, the seed of immortal life, sown in matter by the Ego, has not germinated, and it returns unfruitful and dies—it is an abortion. Many, many seeds never germinate. Many good orthodox, but animal-like lives, live, move, and die,—yes, die in very truth. Would to God I could make all mankind realize this awful, inconceivable privilege of life, that, Jupiter-like, they would turn and face the light.

O child of Adam! "It is easier for a camel to go through the eye of a needle than for a rich man to enter into the Kingdom of God."

PART II

Here beginneth Chapter 3 of the Second Part of the Book which is called "The Tablets of Aeth," wherein the Third, and last, Trinity of the Planetary Rulers is faithfully transcribed.

"Thou hast entered the immeasurable regions. I am the Dweller of the Threshold. What wouldst thou with me? * * * Dost thou fear me? Am I not thy beloved? Is it not for me that thou bast rendered up the delights of thy race? Wouldst thou be wise? Mine is the wisdom of the countless ages. Kiss me, my mortal lover."

"Thus man pursues his weary calling, And wrings the hard life from the sky, While happiness unseen is falling Down from God's bosom silently."

TABLET THE SEVENTH

Saturn

SYMBOL

A human figure with a scepter of power, a being of light crowned with flames.

VII

REFLECTION

TABLET THE SEVENTH

In the external we remember Saturn as an old man, and as a skeleton with a scythe—as Time, in fact. But see, O immortal soul, the real Saturn, as the Angel of Life, having from time gathered the experiences which crown him with light, holding the rod of power; the Christ born in the manger of Capricorn, the Goat—life born of death; the conqueror of evil. He throws off the mask of age, and divine youth beams on us. He doffs the mantle of rags, and royal splendors clothe him. He lifts the hood, and behold the crown. He raises the crutch, and lo! the rod of power. He drops the scythe of death for the jewel of eternal life.

"Om Mani Padme Um." (Oh the jewel in the lotus.)

O child of Adam! Meditate on the transmutations of life. Behold the earthly miracle of the caterpillar and the butterfly, of the toiling mortal and the transcendent God!

TABLET THE EIGHTH

Uranus

SYMBOL

A human eye, from which darts lightning upon an ocean of matter.

VIII

REFLECTION

TABLET THE EIGHTH

The state of soul and spirit—penetration; the wonderful power of soul-perception, which sheds its light on all visible things, receiving their images and interpreting them into the spirit, the all-seer—what does it not convey? The perception that can see deep into your soul and see, as it were, the yet unborn thought; that can distinguish the motive of action; that judges the realities of your soul. Such is the Astral Uranian. For with us all, are three planes of mind: The drift plane, the intellectual, and the spiritual, or internal plane; and thought- reading can be on one or all of these different states. But only the Uranian seer can read the inmost mind, and so really know the possibilities of your spirit.

Imagine an image of soft wax, covered with a sensitive skin. All impressions on the skin shape the plastic wax, but go no deeper— do not reach the soul. You can separate these impressions from your real self, when calm and alone, and look upon emotion as a surface play. But the tragedies of life strike deep. They affect the soul, and go to the center of being. "Verbum sap."

O child of Adam! Watch the tempest of life closely. The Ego may sit calm amidst the storm, but, if that be stirred—BEWARE! The God acts; the soul alone watches.

TABLET THE NINTH

Neptune

A Winged Globe.

IX

REFLECTION

TABLET THE NINTH

An unknown quantity, a hope of progression, ideal love, and all true mental and spiritual ideals; aspiration to become that which we feel to be noble and true; the symbol of the monad, the soul which, receiving its life from the Sun—the Ego—is constantly revealing new forces and potencies of that God-life. Each soul's Ego is its maker and God. The Ego is like the Deific potency of the universe, unlimited in potential power, but limited by its monad as to what will be evolved from its awful depth of being. Deity progresses through its expressions of the cosmos. The Ego, your God, finds progressive expression through you, through your soul. That soul is not immortal that becomes separated from its Ego—its

God. So, soul, spread your spiritual wings and soar upward.

O child of Adam! Know these three things: Eternity is the creator of the universal life; universal life creates the world, and the world is the creator of time. And of these, the Universe is Life, and the World is Mind, and Time is the Soul. The sum total of all is Experience. And this is individual, conscious life—"Jacta est alea" (the die is cast)—the wings are spread.

TABLET THE TENTH

The Cypher - the unknown

SYMBOL

A Shining Nebulae; within it a dot, aimlessly wandering around an unknown center.

X

REFLECTION

TABLET THE TENTH

The unknown in very truth. It is everything—it is also nothing. Inconceivable visions arise within the mental universe, but nothing assumes definite form. It is all that is past. It is likewise everything that the future has in store. Amen.

O child of Adam! "Canst thou bring forth Mazzaroth in his season? or loose the bands of Orion?"

PART III of The Book which is called THE TABLETS OF AETH

OF THE TEN GREAT KABBALISTICAL POWERS or ANGELS OF THE UNIVERSE

PART III

VISION

Each angel standing in front of the symbol is dimly outlined and transparent. Through the angel's form is seen its symbol.

FIRST
A luminous something, which gives the impression of sleep.

SECOND
Something moving, like an ocean.

THIRD
A storm, and lightning.

FOURTH
A mist.

FIFTH
An animal moving, resembling a turtle.

SIXTH
A blue light; in the center a star with three points.

SEVENTH An expanse of water, a blue sky, a shining disk rising on the horizon.

EIGHTH
A lurid sky, like a red dawn; in the water floats an egg.

NINTH Five stars on a convex arc, like a rainbow; the shell of the egg is broken and forms continents.

TENTH A man lying fast asleep under a magnificent palm tree, with his face turned toward the horizon of the sea.

EXPLANATION

Only the pure in heart can see God, and to those pure souls I commend the following brief explanation of the Vision of the Angels of Life, which I have here recorded for the benefit of all whom it may now and hereafter concern.

In the original Vision of the Tablets of Aeth a great circle was seen, in the center a head, a faint shimmer above the head, as if the light were about to dawn; a dull, lurid glow beneath, as if of chaos or hell; the hair around the head like floating clouds, the beard like strange cloud-streaks. Each sign of the Zodiac surrounding the center head had within it a faintly seen face. Beginning with the first, it became more and more distinct and perfect with each sign until it evolved into godlike beauty in Pisces.

The symbolic planets were around the Zodiac, and beyond these, making a third grand circle, were the ten Evolutionary Angels. The vision is that of the evolution of all life, spiritual and material. We gaze at the cosmic sex mystery, and the discerning mind, the loving spirit, can read the correspondence of the great sacred conjugal act of both man and God; of its heights, of its depths, and of all that lies between.

To aid in meditation on the bead at the center, herein is written a vision, an experience of the soul in the Sleep of Sialam.

The Hermetic brethren encircled my astral body, which was deeply entranced. "From whence," the great question, quivered through my inmost being. To answer that awful problem of the soul the released spirit went on its fearsome journey, back through star systems; back, back beyond all stars, back to the blackness of nothing—that awful nothing, whose outside ring vibrated with fearful flames; the fiery cherubim, winged, taking all possible shapes, and unformed living shapes. A human flamed and changed and vanished. The tornado of whirling, flashing, chaotic life swirled and drove through the darkness of chaos of nothing from nothing—and that great, unknown abyss is God! But the life is EVOLUTIONARY.

Deity is progressive, so never can man cease to be. Never can he return to that awful center of nothingness, or be absorbed within the bosom of the unmanifested being. On, and on, and on, with Deific power, God moves in ever-increasing whirls of evolution.

Thus came the answer of the ages: "From primeval force, from the mighty breath of unmanifested being, through every phase of action and reaction, from the energies of storm and lightning, from star-dust to sunlight, has come the spirit of man!"

And the Astral Brethren understood.

THE TWO SEALS OF THE EARTH

I SYMBOL

A human being, with a flaming, burning heart.

II SYMBOL

A round disk inside a light, as from a sun, conceived, but not seen.

So here endeth the Book which is called "The Tablets of Aeth,"
transcribed from the astral originals in the Year of Doom

MDCCCXCIII.
"Omnia Vincit Veritas."

"THY   KINGDOM   COME."
(Zanoni) April, 1893

CHAPTER XII

PENETRALIA

THE SECRET OF THE SOUL

We have now arrived at our final study, which we have approached step by step amid the labyrinth of the mysteries concealed beneath the Veil of Isis. We stand at last upon the very threshold of the sacred Adytum, the "Holy of Holies," from whence proceeds our final revelation of that inmost conception of Man's identity with his Creator—the Penetralia of his Being—the last secret of the incarnated soul.

The written word almost fails us—does fail in fact, when we come to the difficult task of externalizing ideas, the sublimity of which is so infinitely beyond the crystallized images of matter that, they can only be realized in their true glory, when the purified soul can view them from the ineffable heights of eternal spirit. We are lost, dazed, at the brilliancy of the spiritual imagery that opens out before us, in its fathomless stretch of the eternities that are past, of the ever-imperishable present, and the unborn eternities yet to be;

all of them linked together in one grand chain of spiritual relationship and deathless identity; as Man, the Angel, God; and God, the Angel, Man; as the triune Cycle of Being, within the incomprehensible Cycle of Necessity; which constitutes Nature's cosmic university for the complete graduation, education, and purification, of that self-conscious, Deific atom of life, whose expression becomes the human soul. Ah! my brothers could you, but for one single instant, realize WHO you are, WHERE you are journeying, and WHAT your final destiny, every earthly moment at your disposal would be rightly used, and every hour considered too short for your efforts to aid your fellow-man. Selfishness, wealth, and power, would be so utterly contemptible in your sight that their possession would be considered a fearful affliction and a curse, the moment they exceeded the comfortable requirements of mundane existence.

Leave self and the world behind you for the present, and, for the moment, leave your life, with its manifold vanities, in the outer court, and together let us cross the threshold and enter the door of the Temple. There! At last we have entered the Sacred Sanctuary, my brother, and we stand face to face with the imperishable truth of our being—the truth which makes us free, the truth which must ultimately prevail, by virtue of its own inherent Divinity; and we realize Man as he really is, not as he outwardly seems to be. We view him as a molecule, composed of a congregation of separate atoms, all of them held in their places by the centripetal force of the central human atom of life. And yet, small as he is, small as his kingdom is, compared with the mighty creation of which he is a part, he possesses all the inherent qualities of the whole. This, then, is our first conception—Man, is a microcosmic molecule, an atom of divine life.

The scene changes within the chamber, and upon the shimmering, luminous veil, yet before us, we view the large and mighty planet called the Earth. Not as a revolving satellite of the Sun, but as she really is, a vital organ of the macrocosm, the stellar womb of the solar system, the matrix which produces the material organic form of humanity. When the Earth was without form and void," as we are informed in the mystical language of Genesis, the human soul had not yet reached the state, or grade, in the celestial university that desired the Earth and its temporary illusions. Hence this state was void, an unborn idea, the To Be, and darkness, symbolical of complete lack of life and intelligence, "was upon the face of the deep," silent space.

Again the scene changes, and one by one the numberless planets, planetoids, moons, meteors, comets, and other attendant bodies, pass before the eye of the soul as we gaze upon the curtain of this Sacred Penetralia, each orb belonging to some portion of the Astral Man, each great planet constituting some vital function of the macrocosmic organism, and conferring those qualities upon each and every single atom pertaining to that degree of life, so that the solar system becomes individualized as a grand cosmic organism, its attendant satellites constituting its vital organs, and the shining Zodiac its outward form. So, also, each planet is a living, cosmic individual, intensely alive; living, moving, breathing, and bringing forth its offspring of like substance, matter, in obedience to the potential demand of incarnating spirit. The Sun is alive, glowing with life, and constitutes the heart and arterial center of all the circulating fluids of the stellar anatomy.

Scientists may continue to predict, as they have been predicting, the day when solar radiation will cease, but their predictions will prove as worthless as the sighing of the summer wind, so far as reality is concerned. "It is an incomprehensible mystery to science," says Sir Robert Ball, in his "Story of the Heavens," "how the Sun has been able to maintain its heat with such regularity in the past, for there has been no appreciable change in the Earth's temperature for thousands of years." What it is to-day it was ten thousand years ago—yea, Sir Robert Ball, and will be in ten thousand years to come. You may wonder, and the Royal College may wonder, but in the meantime the mighty, pulsating Sun continues beating out its rhythmic vibrations of spiritual and dynamic life—continues, and will continue, to send the exhilarating current throughout every atom, to the remotest part of his solar dominions, and the same current RETURNS TO HIM AGAIN, UNDIMINISHED, for the purification which his glowing, transmuting photosphere alone, can give, to be sent forward again, upon its mission of light, life, and love, around the vital, organic worlds of the astral organism. There is nothing lost, no radiation of energy dispersed upon the unformed, lifeless ether. From the radiating solar focus of Divinity it comes, and to him, undiminished it returns, and so on forever and ever; until the last Deific atom has won its laggard way back to the shining throne of God.

The Sun breathes. The pulsating process of dynamic respiration, eternally repeated during the grand period of a solar lifetime, renews its vital energies, and supplies itself, with the full abundance of the ever-living spirit, transmitted from the shoreless ether in which it lives. It needs no other food, except the magnetic nutriment it receives from each vital organ, or planet, in return for the electrical life current it transmits to them. Just as the human lungs inflate themselves with the vital atmosphere, (which is only the ether, dynamically diluted by the Earth to harmonize with our conditions), to oxygenate the blood and add fresh fuel to the physical furnace, or supply finer essences to the nervous centers. Just as the human heart, with its continual, rhythmic pulsations, propels forth the circulating fluid to every part of the human frame; so does the central heart and lungs of the Grander Man of the Skies, (the Sun) send forth its vitalizing energy to every part of the universe.

Such are the crowding thoughts, born of interior knowledge, that flood the mind as we view these sacred revelations within the sacred chamber of the soul. As yet, we are gazing upon the undulating flow of the astral light. We yearn within our utmost being to become the center of the Penetralia and gaze upon the glorious radiance of the Adonai, from whose ineffable presence we are only screened by the last shining veil of semi-transparent matter, that waves and trembles with every spiritual aspiration. The soul sends forth its pleading cry for light: "Who and what is God?" Faintly, as the distant vesper sounds upon the cooling eve, comes the answer: "Who and what art thou? What canst thou see? What delectable blessing does Nature vouchsafe to the pure in heart?" We tremble with the awful, yet thrilling, revelation. We know dimly, yet fail to realize in our outward consciousness the full import thereof. We realize wherein the mistaken selfhood hath become the only begotten of the Father, but the revelation is too much, and too little. We know that, faint as the voice seemed to the yet unprepared soul, an echo only, IT WAS THE VOICE OF THE ADONAI BEHIND THE VEIL. And now we crave the knowledge of the Where and Whither.

Again, we see the Earth as the vital function of the interplanetary being. It is composed of substance termed matter, which substance is the aggregation of countless atoms, which science has not, and never can, resolve into their individual selves. These atoms are rings of the atomless ether, which, thus differentiated from the formless ether, become centers of force, the center of such force being a vacuum within the atomic ring—a center so small that a microscope with lens one thousand times as powerful as the most perfect modern instrument would fail to reveal it. These atoms form systems, under the control of another apparent vacuum; or, rather, this vacuum seems to be the focus, or center, about which they revolve. THIS SYSTEM CONSTITUTES A SCIENTIFIC MOLECULE OF MATTER, and, in response to the innumerable vibrations, they assume different forms or dimensions, and become, indifferently, molecules of oxygen, hydrogen, nitrogen, or carbon, as the case may be, all of which are but different modes of motion of the same primitive atoms, there being in Nature but three things—Ether, Intelligence, and Motion. What Ether is, no one knows. We call it the formless spirit, the unmanifest, etc. But, there can be no doubt but, that Motion is the product of Intelligence, since we never see Motion but as the manifestation of evolution, and this is the expression of Mind. Therefore, we have a duality—Ether and Intelligence; one the living spirit, the other the eternal substance for its manifestation.

Every molecule of matter is the outward form, the center of which, is the incarnating spirit, in some degree of progress. Man's physical organism is a system of life and development for countless billions of them. So the Earth, in its functional expression as the womb of Nature for the outward expression of Man, is only so in a material sense. HE IS PRIOR TO THE PLANET. He (Man) is only the offspring of the planet by virtue of his material body being a part of the substance of the Earth. This life is a stage, only, of his material journey; and, just as Man's body is continually throwing off useless dead matter and replacing the same with new life, so, too, the countless organic forms of Earth are hourly returning to the ground from which they sprang, and new forms, rising from the same dust, are taking their places.

Here, then, is the sum total: First is revealed to us the grand Astral Man, the Zodiac being the outward idea or form, the Sun and his system the vital functions thereof. The Earth, apart from its functional expression or place, is also an individual. Man, apart from forming a molecule of the planetary womb, by comparison, is also an individual. And, lastly, every molecule of Man's organism is also, in reality, an individual, and small only by comparison with the human frame. And as there are the high Solar Archangels of the Sun, and a chief amongst these seated upon his throne of fire, so there is an Archangelic Chief of the Earth, surrounded by descending degrees of wisdom and power to Man, who also, in his turn, stands as the Deific center and chief of his being, his soul being the sphere of consciousness, which, when united to the feminine soul, constitutes the Angel of Life, Eternal. Down still we go, and find that this Divine scale of life and being is, from the lowly molecule, system upon system climbing, sphere upon sphere, upward and onward, forever, evermore, and all eternity cannot bring nearer the end of Man's glorious immortality.

In the full revelation of this divine scheme of creation, so full of light, life, love, joy and harmony, a scheme void of death and annihilation, the mind once more reflects upon the physical illusions of slowly advancing scientific thought. Camille Flammarion, the great psychomaterialist of France, has painted, in his various novels, a lurid, almost horrible, picture of what the mighty universe must become from the logical deductions of his own school of thought; a school which would be best named as transcendental materialism. According to this conception, "thousands, aye, millions of worlds are rushing through space, inert, frozen, and dead. Suns have cooled down and ceased to give forth the life-sustaining element of light, but have still retained their mighty attraction upon their attendant planets, according to the laws of gravity, by virtue of their material mass, and thus hold their planetary offspring in the eternal, cold, icy grasp of death. Our Sun, too, is cooling fast; the Earth has already lost a great portion of her own internal heat. She has passed her prime of life, and death—cold, icy death—has already begun to encroach upon her extremities. The South pole (the feet) is now practically lifeless in one perpetual covering of ice. So, too, her head; her locks are the white of perpetual snow. No longer has she the blush and beauty of youth, no longer adorned with the healthy covering of verdue which youthfulness gives, and as our geologists prove was once the case. So that, although the time may still be long, according to our reckoning of years, it is only a brief moment in eternity when this fair Earth, and also the beauteous splendor of the silent stars, will be locked forever in darkness, and the final sleep of doom."

If this be so, we ask of the inmost soul, if life be but the fitful awakenings of the indestructible spirit, ebbing and flowing in response to the rise and fall of Nature's cosmic barometer and the transmutations of matter; if life is, in reality, but a brief and passing moment, eternally repeated, from the flush of youth, "the gilded salon to the bier and the shroud, then why, O why should the spirit of mortal be proud?" Why aspire to penetrate the inward realities of life and enter the Holy of Holies— to seek and find out God? As the rushing torrent of this thought swept o'er the mental chambers of the soul and saturated the spirit with its icy sting, as it lay still chained within the prison house of matter, the higher self rose, sublime in its grandeur, and consciousness of divine relationship, and, in the last earthly appeal for light, for divine truth, as to Man and his immortality, it turned in reverence and awe before the still, shimmering veil of the sacred Penetralia. The trial had come, the crucial test, whether of life or death, the final revelation to Man. In purity of heart and humility of soul we await in agonizing suspense. There is a thrilling sensation, as though of ten thousand electric currents consuming the frame, and a swaying to and fro, as if drifting upon an ocean of fire; then a dead silence, so profound that whole eternities seem to pass, without either beginning or end. And the sight of the inward spirit is opened slowly. Who? Where? What? For the shadows have fled, the luminous curtain fades, is gone, and flashing before the inward sight stands the ineffable Adonai. It is I—YOU! There is no God but this, and in one moment the interior consciousness becomes at-one-with-self, God, and from that inconceivable height of profound vision we again look upon Nature. Behold Sun, Moon and planets in all the original magnificence of their nebulous luminosity; from nebulous rings we proceed, stage after stage, each producing its own degrees of life. On, on we pass the ages, the geological cycles of inconceivable duration in time, but only a mere instant in eternity; and on and on, as the changes roll, until we see Earth as she is now; still on, at the ever-urging desire of the triumphant Soul, and a remarkable change is apparent. From forces, at present latent, there comes a change; and, instead of so-called physical; electrical races have superceded the present humanity. Crystallization has ceased; and all things become lighter in density and more ethereal in nature; AND THE ORBIT OF THE EARTH GROWS LESS. Nearer and nearer shines the mighty Sun; first Vulcan, then the swift messenger of the gods are indrawn within the solar vortex, each absorption producing a cataclysmic change upon our Earth. Then comes the turn of Venus, while slowly and surely the orbit of the Earth contracts, and nearer shines the Sun. And, finally, the beautiful Earth, her mission over, the last atom of life beyond her rule, inward she sweeps, and is lost in the mighty ocean of fire as a stone is lost in the lake. Verily is the word of prophecy a literal truth: "The Earth shall be destroyed with fire." And so on with the rest, each planet in its proper turn fulfilling the functions at present performed by the Earth, each becoming the grand theater of material and ethereal life, and the cometary bodies, to-day chasing unknown orbits in the realms of ether, gradually fall into line when their erratic cycle is ended, taking the places of the present outermost planets.

No such thing as death, no such thing as the dark silence of eternal night, for any organic creation of the Most High. From the Sun they come, and unto the Sun each must ultimately return, even as the body of Man, coming from the dust of Earth, must also return thereto, to be taken up in new forms and furnish substance for other degrees of life. And thus will it be, until the Sun, in its mighty solar heavens of purified spiritual life, will form the last, the final battle ground of matter, receiving ITS NEW LIFE FROM A GREATER CENTER THAN ITSELF. A glorious solar world, well typified in the last

Battle of the Gods, and the new Earth—a World whereon the Angels tread in superlatively beautiful forms, clothed with the ideals and emanations of their own divine purity—Souls clothed in Air, treading the ethereal Realms of Light, as the children of God, and the inheritors of the Kingdom of Heaven.

Must the searching eye of the Soul seek further? Must the insatiable thirst of the Spirit launch out upon the trackless infinities of the yet To Be? Must it still penetrate further in the profound beyond, where time ceases to be, where the past, present, and the future, are forever unknown, but exist only as the Deific consciousness of the eternal Now? No. The Soul at last rests satisfied. The final revelation is over.

My brother, we have done; and, in closing, have only to add that, not until the speculating philosophy of earthly schools blends with the Science of the Spheres in the full and perfect fruition of the wisdom of the ages, will Man KNOW and REVERENCE his Creator, and, in the silent Penetralia of his inmost being, respond, in unison with that Angelic Anthem of Life: "We Praise Thee, O God!"

www.ingramcontent.com/pod-product-compliance
Lightning Source LLC
Chambersburg PA
CBHW060842170526
45158CB00001B/211